The Psychology of Fire

❊

DONALD SCOTT

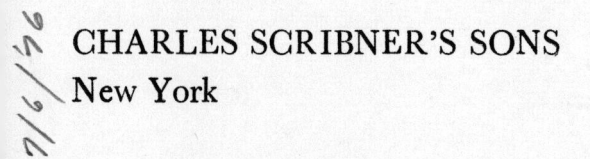

CHARLES SCRIBNER'S SONS
New York

© 1974 Donald Scott

6213

Copyright under the Berne Convention

All rights reserved. No part of this book
may be reproduced in any form without the
permission of Charles Scribner's Sons.

1 3 5 7 9 11 13 15 17 19 I/C . . . .

Printed in Great Britain
Library of Congress Catalog Card Number 74 – 7802

ISBN 0 – 684 – 14015 – 2

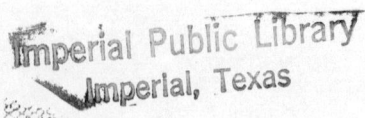

Contents

Preface	vii

PART ONE: THE BACKGROUND

1. The Many Faces of Fire — 3
2. Fire Out of Control — 12
   The great fire of London – The great fire of Chicago – Fire-fighting technology
3. The Destructive Power of Fire — 25
   Aeroplane fires – Fires on board ship – How do fires kill? – The forensic expert
4. Fire-Raising and its Investigation — 33
   The fire investigator – The Menai Straits Bridge fire – Fires in letter boxes – Losses from maliciously set fires – Who determines the cause of fire? – The arsonists

PART TWO: THE MOTIVATED FIRE-SETTERS

5. The Profit Motive — 43
   The Leopold Harris gang – Are there still fires for profit?

6	Political Fire-Raising	55
	The gunpowder plot – The Reichstag fire – Fire as a weapon of war – Aggressive fire – Aggression and depression	
7	Ordeal by Fire	67
	Huss and Palach – Vietnam and suicidal protest – Saints and martyrs – Suicide by fire	

PART THREE : THE MOTIVELESS FIRE-SETTERS

8	The Incendiarists	79
9	Children as Fire-Setters	85
	Sexual Aspects	
10	The Fire-Bugs	93
	Fire-setting sprees – The 'motives' of the fire-bug – Firemen as fire-setters – Case history of a fire-setting fireman – False alarms – Vagrant fire-setters – Women fire-setters	
11	Jonathan Martin, the Incendiarist of York Minster	106
12	Psychotic Fire-Setters	119
	The manic depressive – Alcoholic psychosis – Epilepsy and fire-setting – Other disorders – Legal isues and the problem of treatment	
13	The Fire-Lovers	128
	Sex and fire – Murderous fires – The hijackers – Why fire?	
	Bibliography	139
	Index	145

Preface

Fire was a boon to primitive man and became an element to be harnessed for his benefit. It has made possible much of our civilised way of life, from smelting to electric power, but even in the twentieth century fire is not completely under control. Still it destroys property, and maims and kills people; for as the world becomes more industrialised, more urbanised and overpopulated so these dangers increase. Nor are its effects limited to one or two countries, as even a superficial scrutiny of newspapers will show. A hotel fire in Korea, a dance-hall holocaust in France, an oil-tanker blaze off the coast of South America, a department store burnt to the ground in Belfast; these and many other fires in all parts of the world remind us that it is still a force to be reckoned with.

Much the most sinister aspect of the untamed fire is its abuse by man to cover up his violent crimes, or to make an illicit profit from the insurance companies. Not only this, but it has become a political weapon in the hands of anarchistic individuals – the IRA petrol bombers in Northern Ireland – or of aggressive governments – napalm in South Vietnam. Sometimes the protesting person can use the power of fire to draw attention to his desperation, as did Jan Palach in Czechoslovakia, who met a terrible death – the death meted out to saints and witches in the Middle Ages. The abuse of fire takes yet other forms. It is used by the 'fire-bug' for his own disordered pleasure as well as to bring havoc on society. And it is employed by some as a tool to wreak vengeance on those who have insulted them, either in reality or in fantasy, thus becoming part of the deranged mental processes of the psychotic person.

The aim of this book is to show just how uncontrolled fire really is; how it is an increasing menace and, as it were, a source of pollution; how some men can and do use and abuse fire to their own ends. Much of the information on this topic is not readily available, either to the professional or to the general reader. It is hoped that this account will add to our present stock of knowledge on the subject, as well as raising the question of what can be done. As a psychiatrist the author has experience of people who abuse fire, experience which has helped to sharpen his interest in the field so that he has come to understand more clearly some of the motives of the fire-raiser; it is information which may be helpful not only in the taming of fire but in our understanding of the deranged or psychotic mind.

For those who wish to follow up any of the points raised, a detailed bibliography of sources is provided.

Acknowledgments

I wish to thank the many colleagues who helped by providing information on the various aspects of fire covered in this book, in particular Miss I. Binnar of the Fire Protection Association, Mr E. D. Chambers of the Fire Research Department of the Department of Environment, Mr Frederick R. Hyde Chambers, General Secretary of the Buddhist Society, Dr W. R. James of the Institute of Pathology at the Welsh National School of Medicine, Professor T. C. N. Gibbens of the Institute of Psychiatry and Mr Richard Sear of the *Daily Mirror*. I am particularly grateful to the Chief Librarians and their staffs of various libraries who have located suitable material, including the Blackheath Public Library, the British Medical Association, the *Daily Mirror*, the Fire Research Department, the Institute of Psychiatry and the London Hospital Medical College.

In addition I am indebted to Mrs Joan Butcher and Mrs Jean Held for secretarial work during the preparation of various drafts of the manuscript and to A.M.S., to whom this book is dedicated, for her assistance throughout.

For quotations from copyright works I make acknowledgment to the following: for the passage from 'Little Gidding' in *Four Quartets* by T. S. Eliot to Faber & Faber Ltd and Harcourt Brace Jovanovich Inc; for 'Mrs Beaumont' in *Daybreak* by Joan Baez to Dial Press Inc. and Fontana Books Ltd; for sections from Thomas Balston's *Life of Jonathan Martin* to Macmillan & Co. Ltd; for the poem 'Fire and Ice' by Robert Frost to Jonathan Cape Ltd and Holt, Rinehart & Winston.

London 1974   D.F.S.

PART ONE

THE BACKGROUND

Chapter One

The Many Faces of Fire

And he tamed fire which, like some beast of prey,
Most terrible, but lovely, played beneath
The frown of man; and tortured to his will
Iron and gold, the stars and signs of power.
                    Percy Bysshe Shelley, *Prometheus Unbound*

Fire is an essential element in our world. Without the sun – a ball of nuclear fire – we would have no heat, no light and, indeed, would not exist, because heat is necessary not only for the continuance of life but for its very creation. However, for our present purposes it is not so much the existence of fire as the harnessing of it which is important, to prevent its destructive power being unleashed against forests, buildings and the inhabitants of our bustling cities. Clearly in urban communities the control of fire is of very great importance; nevertheless, at the present time it is the highly industrialised societies which, in spite of improved fire-fighting technology, show a marked increase in numbers of conflagrations.

We do not need to ask how fire came to this planet, because it was always here. The earth started as a ball of fire, separated from some other heavenly body; gradually, as it cooled, life became possible on the surface. How did primitive man first come to discover and tame fire? Imagine him seeing a volcano in the distance, or the lightning during a thunderstorm, perhaps striking a tree, which was then set ablaze. It is difficult to know how he would have reacted. Would he approach closer out of curiosity, or would he run away? At what stage did he learn not only that fire was hot but also that he himself could capture it and subsequently produce it at will?

From observations of primitive tribes who inhabit remote parts of the world at the present time, it is known just how fire was produced by those early inhabitants of our earth. The Tasaday people of southern Mindanao in the Philippines are one example of a primitive tribe who make fire by twirling a dowel-like drill in a hole made in another piece of wood. They then use dry, mossy tinder to catch the first sparks and gently blow it until a small flickering flame appears. It is a time-consuming process, and we can appreciate how primitive peoples today, as in the past, are keen to conserve fire rather than continually have to rekindle it. Thus they carry it with them in a hollow stick and relight their fire when and where it is needed for cooking or warmth.

Because of the power and the value of fire, primitive peoples were bound to create stories about how it came to earth; a subject discussed in depth by Frazer in *Myths of the Origin of Fire*. As civilisation progressed, these legends became more and more involved. Perhaps the legend of Prometheus is the most complex and interesting of all. Prometheus, the Greek Titan, is credited as the founder of civilisation because he stole fire from the gods, but as a result of his trickery he was chained to a rock where birds fed on his liver. This legend has been studied in great detail by the psychoanalysts Freud and Jung, who consider that it is not only of interest in itself but also helps to explain various traits in those who misuse fire. Prometheus was an aggressive male and thus felt the need to outwit the gods and capture their fire. He carried it in a hollow stick, which can be seen to be at the same time a hollow or female symbol and yet externally phallic and male. This implies some confusion of sexual identity, and certainly, as we shall see later, people who misuse fire do have poor sexual adjustment.

However, it is not only the act of stealing fire and bringing it to earth, but the eventual punishment of the person who brought it, which is important. Prometheus was fastened to a boulder and forced to accept a completely passive role as the vultures tore at his entrails. This too is intriguing, since again it has implications for the fire-raiser, who after his aggressive act stands back, accepting a passive role as the fire takes over; licking tongues of flame also convey a female-passive aspect of fire, another feature noted by Jung.

The Prometheus legend is not the only one concerning fire in Greek mythology. Another relates the story of a father-son relationship. Phaethon, which means 'shining' or 'radiant', was the name of the son of Helios the Sun God and the nymph Clymene. There was some doubt about Phaethon's legitimacy and in order to prove himself he asked his father if he could drive the Chariot of the Sun for the day. His father begged him not to do so, but Phaethon insisted and proceeded to drive the chariot too close to the earth so that the ground was scorched. Zeus became alarmed in case further damage occurred and hurled a thunderbolt at the chariot. Phaethon fell to earth and was drowned in the river Po.

For most people fire has its pleasurable attributes. The joy of the open grate with its play of light and colour is without equal as a releaser of fantasy. Indeed, the fireside is a marvellous place for reverie, even more so perhaps when the wind is howling outside, dashing the inhospitable rain or snow against the window. The fireside, too, is a place for companionship, and for intimate conversation.

In *The Blood of Others* Simone de Beauvoir, clearly aware of the intimacy of the fireside, sees that fire like love has an all-embracing quality. And John Wilmot, the Earl of Rochester, draws the same parallel, but with typical seventeenth-century bluntness and humour :

> Have you not in a Chimney seen
> A sullen faggot, wet and green,
> How coyly it receives the heat,
> And at both ends does fume and sweat?
>
> So fares it with the harmless Maid
> When first upon her Back she's laid
> But the kind experienced Dame
> Cracks, and rejoices in the Flame.

Fire is a springboard for humour as well as evoking sensual and sexual pleasures. It was seen by St Francis thus : 'Brother fire, fair, jocund and most robust and strong.' And Harry Graham was similarly struck by the jollity of fire, though he mixed the wit in his verse with warnings of that other side of its nature:

> Billy, in one of his nice new sashes,
> Fell in the fire and was burnt to ashes;
> Now, although the room grows chilly,
> I haven't the heart to poke poor Billy.

The warmth of fire and the pleasure it gives are not its only positive attributes, for it cooks the food of both the hunter and the monarch. Who first came to utilise fire for cooking is of course totally unknown, but Lamb in his *Essays of Elia* has written a playful fantasy on the discovery of roast pork. The story tells of a farmer whose pigsty is burnt down. Struck by the delicious smell, and touching the roasted flesh of the unfortunate animals, the farmer rapidly comes to appreciate the delights of roast pork.

The ancient Greek physician Hippocrates was well aware of the beneficial effects of cauterisation. He wrote: 'What medicine does not cure the iron does, and what the iron does not cure fire exterminates.' This use of fire must be well known to the addicts of Western films; in the West a small amount of gunpowder was often poured into a wound and then ignited – a primitive form of cauterisation. An even more devastating and dreadful technique, the application of a red hot iron, was used by surgeons in the Middle Ages and later. In his *Apologie and Treatise* Ambrose Paré recounts the disadvantages of this form of therapy:

> Now so it is, that one cannot apply hot irons but with extreame and vehement paine in a sensible part.... Moreover, it would bee a long while afterwards before the poore patients were cured, because that by the action of fire there was made an eschar, which proceeds from the subject flesh, which being fallen, nature must regenerate a new flesh instead of that which hath been burnt, as also the bones remaines discovered and bare; and by this meanes from the most part there remaines an ulcer incurable.

It is easy to understand from these observations that many sick people suffered severe pain at the hands of their doctors, pain which probably brought little benefit since the underlying condition in most cases either would have been lethal or would have resolved itself in time. When eventually doctors and surgeons learnt to do the least possible damage to their patients and certainly not to inflict extra pain and suffering, a great advance had been achieved.

What was the basis of this treatment by cauterisation? One might imagine that inflicting another wound in addition to the one already there would cause even more devitalisation of tissue, which itself would form a food substance for the destructive bacterial agents, this in turn leading not only to severe local infection but to widespread, blood-borne septicaemia. Ambrose Paré was well aware of the difficulties and he decided to try something else. He writes: '[This] occasion moved the author to devise this new forme of remedy, to staunch the blood after amputation of the member; and forsake the common way used almost by all chirurgions; which is, by appplication of actuall cauteries.' His technique was to bind the stump, thus staunching the flow of blood; he had observed that cauterisation by itself could not stop the flow of blood and that the method led not only to severe pain but to fever and convulsions as the infection gained an even greater hold.

Much later, in the late nineteenth century, fire was used in the sterilisation of surgical instruments, after an initial period of antiseptic surgery when chemical agents were mainly employed. Even today boiling or steam are the common means of sterilisation in operating theatres, though atomic radiations have been employed for disposable equipment such as syringes and needles.

Before fire was known to be a true cleansing agent, it had a symbolic role in this respect. Just how this came about is uncertain, though perhaps it was through an analogy with the purification of metals. For example, when gold is heated the impurities rise to the surface and can be removed. It is a technique which seems to be reflected in the phrase 'I have passed through the fire'. Like gold, the person is presumed to have been purged of all the dross. Shakespeare touches on this in *Twelfth Night*. 'You should have accosted her; and with some excellent jests fire-new from the mint you should have banged the youth into dumbness.'

The expression 'I will go through fire and water', however, has a different origin. It refers to the 'trial by ordeal', a common method of trial in Anglo-Saxon England. The witch, or it may be some other offender, would be thrown into the river. If he or she survived this part of the ordeal, the next task would be to carry a hot ball for a specified distance.

Fire was also used by ancient Greeks (generally thought of as civilised) in a surprisingly vicious way. Citizens who committed suicide were burnt and then buried outside the city; the hand, which had committed the frightening act of self-destruction, was cut off and buried separately from the other remains.

Even today fire is still used in ceremonies of purification, notably by Buddhists in the fire-walking ceremony. Originally a Hindu custom, it was taken over by the Buddhists and forms part of the autumn ritual for members of this religion in India and Ceylon. On the evening before the fire-walking, special wood logs are kindled in a long pit in front of the temple. The fire burns all night and in the morning the hot embers are all that remain, the temperature of the whitish-looking ashes being about 600°F. The walkers gather at one end of the pit, standing in a special liquid mud before they walk across the embers. They are rarely burnt, and if there are signs of this it is not on the soles of the feet but on the legs or insteps. That they are not burnt is taken to indicate their purity, in the same way as the person who was not seared by the hot ball in primitive Anglo-Saxon times was judged to be innocent. The faces of the fire-walkers show a serene, almost heavenly, expression, certainly not a look of pain and anguish.

Fire has also come to be equated with life. In Roman times the sacred fire was entrusted to the Vestal Virgins, whose task it was to keep the light burning always. The flickering flame in the Arc de Triomphe in Paris represents the continuation of the spirit of men killed in the First World War; it tells the world that they are not forgotten. Fire is used in another sense in the following lines by W. S. Landor:

> I warmed both hands before the fire of life,
> It sinks and I am ready to depart.

In the Catholic Church too there is the ceremony of 'the making of the new fire', part of the liturgy which leads up to Easter Sunday. This particular ritual emphasises not only the purificatory aspects of fire, but also its qualities of rebirth and renewal.

But these are not the only religious aspects of fire, for in both the Hebrew and the Christian ethic fire is an agent not only of cleansing but of sacrifice and punishment, this latter association coming to assume catastrophic proportions in the overwhelmingly destructive vision of hell, perhaps one of the earliest and most terrifying aspects of fire that children of our culture have become aware of over the centuries. The impact of this association is examined by Lauretta Bender in her writings on fire-raising by children, an author to whom we shall return in subsequent chapters.

Already we have seen that fire, in both its benign and its malignant aspects, touches on man's life at every turn. Indeed, it crops up endlessly in our everyday speech. We talk of burning our fingers, burning the candle at both ends, being placed between two fires, burning our boats. New ideas are blazed abroad; revolutionaries are firebrands, enthusiastic colleagues are balls of fire, new girl-friends are hot numbers. Examples could be produced almost endlessly. Fire, too, symbolised for T. S. Eliot the all-consuming passion of divine love, when he wrote in *Little Gidding*:

> Who then devised the torment? Love.
> Love is the unfamiliar Name
> Behind the hands that wove
> The intolerable shirt of flame
> Which human power cannot remove.
> We only live, only suspire
> Consumed by either fire or fire.

But these are perhaps adult sophisticated views of fire. How does our experience of it begin?

The baby sits fascinated in front of the open fire. But he soon learns that it is the one thing he cannot touch. (This is assuming, of course, that his mental development is normal, for the severely mentally handicapped person never understands the dangers of fire, in the same way that he never understands how to cross a road safely). But as our normal child reaches the age where he is allowed to play with other children in the area, he is likely to wander on to sites where buildings have been demolished or to disused parking areas. Here, together with his young friends, one of whom may have a box of matches in his pocket surreptitiously filched from the kitchen, it may be decided to start a fire with the rubbish lying around. They know that they may get burnt and that the flames may get out of control, but they enjoy the warmth and the spectacle just as they delight in the organised bonfire activities on 5 November, Guy Fawkes night, the celebration of a shadowy figure of whom we shall hear more later.

Gaston Bachélard in his *The Psychoanalysis of Fire* writes about these same early experiments with fire, though the children he envisages are country-born and have, perhaps, more scope for their activities. Disappearing with a companion over the fields, they find some small dell where, with their stolen matches, they can kindle a fire; thus it is that they recapitulate the Prometheus legend, taking fire from the gods who are their parents. In this secret place the boys warm themselves and enjoy hot potatoes cooked in the embers, perhaps scorched and scalding on the outside but hard and uncooked in the middle.

Partly through their own common sense and partly through the advice of their parents, most children soon adjust to the necessary restrictions on their behaviour, not only with fire. However, some children apparently do not learn; as they grow up they become the fire-bugs, the arsonists, the incendiarists about which we read, during their schooldays setting the classrooms on fire, in their adolescent days climbing into a woodyard or a factory and starting a blaze there, in their adult lives setting fire to farm-buildings, hospitals or churches. Some of the reasons why they do will be examined in this book.

Twentieth-century fire has lost much of its former liberty; it is now largely captured. Now the open fire is almost a thing of the past, the flames enclosed within some 'central heating' system. Fire is in fact so controlled that it is at the beck and call of man when he wishes to light a cigarette; it has become something as trivial as that. But let us not become complacent. Fire can still get out of control only too easily; and a final vast conflagration is still anticipated by many, as Robert Frost points out in his poem 'Fire and Ice':

Some say the world will end in fire,
Some say in ice.
From what I've tasted of desire
I hold with those who favour fire.
But if it had to perish twice,
I think I know enough of hate
To say that for destruction ice
Is also great
And would suffice.

Chapter Two

Fire Out of Control

> If men could learn from history, what lessons it might teach us. But passion and party blind our eyes, and the light which experience gives is a lantern on the stern which shines only on the waves behind us.
>
> T. Allsops, *Recollections of Coleridge*

Even in the centuries before Christ, fire had begun to serve men well, but almost from the moment they began to harness it, fire was likely to get out of control. As now, this was an even bigger problem in urban communities than in the country. There was an enormous conflagration in Rome, allegedly started by Nero who is also said to have taken particular pleasure in the blaze that he created. Tradition has it that he 'fiddled as Rome burnt'. But his glee, according to the historians, was based on his wish to clear an area housing the 'lower classes', so that he could further his grandiose plans for a great palace.

The magnitude of fires and their frequency probably reached a peak in the medieval cities of Western Europe. These had narrow streets, the houses were built largely of wood, and because there was no rubbish disposal this combustible material collected in the narrow thoroughfares and up against the houses. Then, as now, rubbish was often accidentally and sometimes maliciously ignited.

Constantinople, according to Frank Lord, was ravaged by fire in the eighteenth century. In 1729 it is said that twelve thousand houses were burnt and seven thousand people perished. Twenty years later another fire killed an even larger number of people, the figure being estimated at about ten thousand. Some of the greatest fires the world has known have followed earthquakes, like that of San Francisco in 1906 in which seven square miles of the city were devastated, a thousand lives lost and an estimated $120 million worth of damage done. But it is the fire of London which is of particular interest to us for it showed that 'man can learn from history'; to plagiarise Coleridge it indicated that the light can shine not only on the waves astern but also on those ahead.

The great fire of London
In 1665 London was ravaged by the plague. This, like the fire, was another of the terrors of medieval life; always sparking and flickering somewhere, suddenly it would flare up. This particular outbreak of plague, though it was severe in London and its environs, was by no means the worst that had struck the country. Nevertheless there were many deaths; how many is uncertain but perhaps a third of the population of the City of London perished.

Curiously enough, during the plague the medical uses of fire were of great importance. Defoe, in his *Journal of the Plague Year*, says that after the plague the infected houses, apparel, bedding and hangings of chambers were 'well aired with fire and such perfumes as are requisite within the infected house before they can be taken again into use. Apparently a great variety of substances were burnt for this purpose of fumigation, including gunpowder, pitch and sulphur. Fire was also used for cauterisation. The plague led to extremely painful swellings in the groin and under the arms and surgeons called in to treat these swellings used strong caustics or cauterised the flesh with hot irons, thus hoping to burst the swelling and allow the suppuration to escape.

Fumigation and treatment were two uses of fire; but this was not all, it was supposed to have a preventive action in that the noxious vapours and particles emitted by coal fires could destroy the infection, thus keeping the air in the room sweet and clean. This idea that infection is spread by the air was still prevalent in the nineteenth century; indeed Lister, the father of modern surgery, regarded the air as a source of contamination and sprayed carbolic acid around the site of a surgical operation to purify it. But to return to the seventeenth century. A certain Doctor Hodges of that period was so convinced of the preventative properties of fire that he suggested to the authorities that bonfires should be lit in the streets to control the spread of the plague. After much disagreement and argument, a trial of such a measure was decided upon and carried out, but it was not regarded as successful and was therefore not generally adopted.

It is interesting that Defoe observed the similarity between the plague and the fire. He wrote:

> The plague, like a great fire, if a few houses only are contiguous where it happens, can only burn a few houses; or if it begins in a single, or, as we shall call it, a lone house, can only burn that lone house where it begins. But if it begins in a close-built town or city and gets ahead, there its fury increases, it rages over the whole place, and consumes all it can within reach.

It was less than nine months after the plague had subsided that the City of London was purged with flames and was brought to ashes. The plague killed an estimated fifty-six thousand people; the fire was remarkable in that there were few recorded deaths attributable to it, something that could not be said of the inferno that raged in Chicago in 1871 when three hundred people died.

There are few other incidents in London's history that have been more significant than the great fire, in which the flames raged through the streets for four September days in the year 1666, destroying utterly the ecclesiastical and monastical city. A new city was to rise in its place. Samuel Pepys, the famous diarist, who at the time was living in the Navy Office at Seething Lane not far from the Tower of London, was told of the fire when it first began but thought little of it. Like the lord mayor of London, he was not to know that from its necessarily small beginnings it would eventually destroy virtually the whole of the medieval conurbation, the London that Shakespeare knew and that the city companies had built up.

The fire, like many great events, had been predicted. Indeed the king himself had expressed his fears and the year before had written to the lord mayor warning him of the danger of fire because of the narrowness of the streets and alleys and the fact that the overhanging houses were built of wood. As Bell in *The Great Fire of London* puts it, 'through the early years of the Restoration, there runs a vein of gloomy prognostication of the impending catastrophe to be brought about by God's vengeance upon sinful London'. And in a rather biblical vein, Walter Gostello wrote: 'London, go on still in thy presumptuous wickedness! Put the evil day far from thee, and repent not! Do so, London. But if Fire makes not ashes of the City, and thy bones also, conclude me a liar for ever. Oh, London! London! Sinful Sodom and Gomorrah! The decree is gone out, repent, or burn as Sodom and Gomorrah!' That was in 1658, eight years before the fire.

Where did it start? In the shop and house of one Farynor, the king's baker, whose premises were in Pudding Lane. As the story has it, by two o'clock in the morning on 2 September 1666 the flames had broken out in his oven, which he had liberally stoked before retiring to bed. Later we shall hear of the accusations about Papists or Frenchmen who set the city on fire; in the meantime Farynor, his day's trading over, was asleep, a pile of firewood ready by the side of his oven so that he could get it going in the morning. The morning was to begin earlier than he had expected. At two o'clock he was awoken by choking fumes. The fire was as yet not very fierce, but he roused his family and they escaped through a garret window into a neighbour's house. The servant who had given the warning also made his getaway but a maidservant, more timid than the others, stayed behind and became the first of the few victims of the conflagration.

Of course fires were not unusual in London, as in many other large cities at the time, and so the blazing backhouse was of no particular importance. The lord mayor of London, Sir Thomas Bloodworth, when he first surveyed the flames said, 'Pish! A woman might piss it out'; nevertheless, after only one day of the holocaust he disappeared – he could not cope with the situation. Quickly the church of St Magnus the Martyr was burnt; London Bridge lay directly in the path of the blaze.

Samuel Wiseman wrote (cited by Bell):

> And now the doleful dreadful hideous note
> Of fire, is screamed out with a deep strained throat;
> Horror, and fear, and sad distracted cries
> Chide sloth away and bid the slugged rise;
> Most direful acclamations are let flye
> From every tongue, tears stand in every eye.

The wind was an important factor: it blew and blew and fanned the flames so that the fire extended and expanded for four days. As a result of the wind's constant strength and direction the city lay defenceless before the blaze. What was to be done? Little. There were leathern buckets, ladders, axes and stout hooks but what use were they against the acres of roaring flames whipped up by the wind? This uselessness of the fire-fighting equipment was to be the stimulus for the development of really effective appliances, one of the first results of the fire of London.

In any case, the equipment was dispersed in city churches under the towers and in the livery company halls. Another problem was where to get the water to fill the heavy buckets. There was a primitive apparatus at London Bridge to raise water from the Thames but the system of pipes and pumps was quite inadequate. The fire engine had been invented in 1651, one capable of throwing jets of water on to the flames, but this was quite unknown in England in 1666. There were certainly hand syringes made of brass available during the seventeenth century, but even these do not seem to have come into prominence until after the fire of London. It is interesting that Sir John Robinson, lieutenant of the Tower of London, obtained authority to provide 'four hundred engines, three hundred buckets, ten ladders and twelve hooks for the suppression of fire in the Tower of London should it occur'. The authority was dated 8 November 1666, two months after the destruction of the city.

The city after the first day of the fire was a picture of hopeless confusion. Pepys gives an account thus:

> ... everybody endeavouring to remove their goods, and flinging them into the river, or bringing them into lighters that lay off; poor people staying in their houses as long as till the very fire touched them and then running into the boats, or clambering from one pair of stairs by the waterside to another ... having stayed and in an hours' time seen the fire rage everywhere, and nobody, in my sight, endeavouring to quench it, but to remove their goods and leave all to the fire; and having seen it get as far as the steelyard, and the wind mighty high and driving it into the city; and everything, after so long a drought, proving combustible, even the very stones of the churches. ... I go to

Whitehall, with a gentleman with me, who desired to go off from the Tower, to see the fire in my boat.

It was still some time before effective action began to be taken to clear buildings from the path of the fire, so that it could be contained. As the fire progressed the excitement spread; rumours abounded and bands of the homeless swarmed through the streets clutching their hastily snatched belongings and attempting to find a place of safety from the advancing flames. In this atmosphere of crisis, apprehension grew and there were stories of Frenchmen assailed by the mob. It was difficult, the rumours went, to accept such a conflagration as an accident; surely it must be a plot of the Dutch, the French, the Papists, to destroy the city. Bell quotes as follows: 'Sir Edward Southcote, passing to his country seat through London while the city was still burning, had one of his servants seized, a Frenchman, a man being overheard to speak broken English, but by explanations and a gift of half a crown to the captors for drink obtained his release.'

Bell also writes of Malcolm's vivid description, in *Londinium Redivivum*, of the people fleeing, carrying on their backs whatever they could take. They were leaving a city which seemed doomed entirely.

> The flames were all in the wind, and to the sights and sounds of those awful hours was added the sharp crack of powder explosions, followed by the rattle and thud with which a shattered house fell to the ground. Those who looked back into the red heart of the flames saw houses wreathed in flame and swaying, then toppling headlong into the roadway. The fire sent up a huge volume of smoke as from some titan's furnace, streaking with flying sparks . . . the sun shone red . . . yielding a fainter light than in an eclipse. . . .

W. H. Ainsworth in *Old St Pauls* gives a graphic account of the destruction of St Paul's Cathedral.

> Flames were likewise bursting from the belfrey, and from the lofty pointed windows below it, flickering and playing round the hoary buttresses and disturbing the numerous Jackdaws that built in their time-worn crevices, and now flew screaming forth . . . It now became evident also, from the strange roaring noise proceeding from the tower, that the flames were descending the spiral staircase, and forcing their way through some secret doors or

passages to the roof. . . . The cries of the multitude, coupled with the roaring of the conflagration, resounded from without, while the fierce glare of the flames lighted up the painted windows at the head of the choir with unwonted splendour. Overhead was heard a hollow rumbling noise like that of distant thunder, which continued for a short time, while fluid streams of smoke crept through the mighty rafters of the roof, and gradually filled the whole interior of the fabric with vapour. Suddenly a tremendous cracking was heard, as if the whole pile were tumbling in pieces. . . . The flames now raged with a fierceness wholly inconceivable, considering the material they had to work upon. The molten lead poured down in torrents, and not merely flooded the whole interior of the fabric but ran down in a wide and boiling stream almost as far as the Thames, consuming everything in its way and rendering the very pavements red hot.

So the conflagration continued until the whole Cathedral was a mere shell and, in the words of John Evelyn, the other famous diarist of the time, 'One of the most ancient pieces of piety in the Christian world was a heap of ruin and ashes.'

Only a Nero could have enjoyed the spectacle. At the time the magistrates cast foreign residents into prison, apparently because they were suspected of the crime but in fact to protect them from the fury of the populace – they were afterwards released. Indeed all the outlying jails were full during the period of the great fire of London. There was even a suspicion that the first sparks had come from heaven, as a punishment for the evil of London.

The aftermath of the fire is well described by Evelyn. He had helped to keep Holborn from the flames and he describes what he saw on Friday, 7 September:

> I went this morning on foot from Whitehall as far as London Bridge, through the late Fleet Street, Ludgate Hill by St Paul's, Cheapside, Exchange, Bishopsgate, Aldersgate and out to Moorfields then through Cornhill etc. with extraordinary difficulty, climbing over heaps of yet smoking rubbish, and frequently mistaking where I was: the ground under my feet so hot that it even burnt the soles of my shoes.

He goes on to report how St Paul's was a ruin, how many city companies' splendid buildings were reduced to dust, fountains dried out and ruined, stones whitened to the colour of snow, and clouds of smoke everywhere. He continues: 'In the midst of all this calamity and confusion there was, I know not how, an alarm begun that the French and the Dutch, with whom we were now in hostility, were not only landed but even entering the city.'

It was not only the Dutch and the French who were blamed; the Anabaptists, Fanatics, Nonconformists, Quakers and even King Charles II himself were said to have been part of a plot to burn the city; but it was mostly the Catholics who were accused.

There was also a tale that a ten-year-old boy, Edward Taylor, had with his father set light to Pudding Lane. Apparently, the story went, they had taken two fireballs made of gunpowder and brimstone, lit them, 'then flung them into the window so setting the house alight; this was at the beginning of the firing of London'. However, Lord Lovelace was unable to obtain any tangible evidence of such a story. Nevertheless in the excitement (and fire always causes individual and corporate panic) various people were put on trial at the Old Bailey for starting the blaze; but no one at this stage was found guilty.

Parliament next set up a committee under the chairmanship of Sir Robert Brook to enquire into the causes of the fire. They presented their report to Parliament on 22 January 1667; in fact nothing came of the report. However, the citizens of London were not satisfied and they still felt that the Papists were to blame. A real scapegoat had to be found.

'The occasion demanded a victim and the excited populus, inflamed with a hostility to foreigners and Papists and distressed by their losses, were little inclined to reason coldly': thus the words of Bell in *The Great Fire of London*. A victim was duly found. He was a Frenchman called Hubert, apprehended in Romford apparently trying to escape from the country; a watchmaker by trade, a native of Rouen and twenty-six years of age. The story of Hubert varied with the telling, but apparently with a certain Stephen Peidloe he left France, sailed to Sweden, and from there to England in a Swedish ship called the *Skipper*. This ship was moored in the Thames near St Catherine's Tower. Hubert and Peidloe stayed on board until the evening of the fire. Then the two went ashore to Pudding Lane and reached Farynor's the bakers. Peidloe, it is reported, said to Hubert that he had fireballs with him, and he gave Hubert one to throw into the house. Hubert said that he placed the fireball on a long pole, lit it and thrust it through the window, staying until the house was well alight. The strange thing is that the window into which Hubert said he pushed the blazing torch did not exist, a fact confirmed not only by the baker but by his son and daughter.

Hubert was tried at the Old Bailey in the October sessions, where further contradictions were noted at the trial. There it was said that he had come in company with many other assistants and with a clear plot to set fire to the city. His reward for the task was alleged to be merely a pistol.

The testimony of a French merchant living in St Mary Axe corroborated the fact that Peidloe (who was never seen at the trial) was a wicked man and that Hubert had since childhood been a waster. Hubert confessed that he was guilty. Strangely, he said that he was a Catholic and yet it was well known that he was a Protestant; it further transpired, from the testimony of the master of the *Skipper*, that Hubert had not gone ashore until two days after the fire had begun in the baker's house. In spite of all this, Hubert was found guilty and hanged at Tyburn. 'He died, it was said, reasserting his guilt.' Clarendon, Earl Edward Hide, in his 'Life' observes of Hubert, 'though no man could imagine any reason why a man should so desperately throw away his life, which he might had saved though he had been guilty, since he was only accused upon his own confession; yet neither the judges nor any present at the trial did believe him guilty, but that he was a poor distracted wretch, weary of his life, and chose to part with it in this way.'

Hubert, it seems certain, was not guilty of the act for which he was hanged. However, the whole affair shows certain important features relevant to our theme. There seems to be little doubt but that he was mentally disturbed, confessing as he did to a crime he had not committed; this is something still well known even today for a variety of crimes including arson and murder. The fact that the populace had found a scapegoat who was a foreigner and said to be a Roman Catholic is also important, for it shows how minorities are oppressed and blamed even for events in which they have no part. Perhaps there is a parallel here in the political climate of Northern Ireland where, as we shall see later, arson is so common, destructive and costly that the whole economy is in danger of collapse.

Superstition has it that the fire of London scorched the ground and finally killed the plague. This may or may not be true; what is undeniable is that the fire, in spite of its devastation, brought about some important changes in many directions. It certainly destroyed the decaying, insanitary houses in the narrow streets, and that the London which arose after the conflagration was a different London is not questioned; it was one which, although still evil and squalid, was never as unwholesome as it had been when the timber-framed houses, with gables practically meeting across the narrow streets, prevented the light of the sun reaching the thoroughfares.

The extent of the destruction can be understood when it is realised that St Paul's Cathedral and eighty-nine churches were razed, thus allowing Wren as well as other architects, in a spate of unsurpassed energy, to rebuild. The new St Paul's was constructed of white Portland stone and the houses which were built after the devastation were of red brick rather than wood and plaster.

There were two other even more far-reaching results. The first was the setting-up of fire-fighting services, under the auspices of the insurance companies (see below). Later, in the nineteenth century, after a conflagration in Tooley Street, these services were amalgamated to form the predecessor of the force now called the London Fire Brigade. Another feature was the development of effective equipment. Even before this time, as we have mentioned, some primitive fire engines had been devised. It was Yan van de Heiden together with his son Yan who published the earliest book on fire engines and these two were instrumental in devising hoses made of leather which could be directed more easily at the flames than the swivelling goose-neck delivery jets which had been used before. The engines themselves were pumped by hand and these too had flexible hoses which could draw water from the canal or river bank.

One further result was the beginning of fire insurance. The first companies were set up in London and each office had its own fire mark. These were put on the fronts of their clients' houses or premises, so that if there was an outbreak the fire equipment of that particular insurance company could be rushed to the spot. The Phoenix was one of the first companies, formed in 1682, its famous fire mark showing the legendary bird rising from the flames. Another was the Salamander fire mark, showing a lizard-like creature which was supposed to dwell in a world of flames. The Guardian, another early fire insurance office, used Minerva the Goddess of Wisdom for its sign. These tangible results of the fire were not limited to England, but were eventually to have a worldwide influence on the control of fire.

The great fire of Chicago

Fires continued to rage, as they still do today in residential and industrial areas, for big fires are a product of an industrialised community. Chicago in 1871 was a rapidly expanding city and was to become the scene of an immense conflagration. Its expansion had gone on since the middle of the century, and the rapidly increasing population produced difficulties in housing so that the buildings were largely constructed of wood. It is estimated that in the blaze 17,450 houses were destroyed and 98,000 people were rendered homeless. In the fire of Chicago three hundred people lost their lives, mainly because of the problem of escaping from the rapid spread of the flames. The streets were full of struggling mobs, all was utter confusion, and the wooden houses, dry because there had been little rain over the past months, collapsed round the struggling people.

The fire started on Sunday, 8 October 1871, and raged until 4 am on the Tuesday, when by great fortune heavy rain helped to extinguish the blaze. The burnt-out area comprised 2,124 acres and, intriguingly, there were only two houses standing in that whole expanse, the mansion of Mahlon Ogden and the modest home of Richard Bellinger, a policeman. Bellinger raked the leaves from his garden and burnt down his wooden fences, keeping the roof of his house wet, first by using water from a cistern and then by carrying buckets from a ditch two streets away. Finally it is said he poured his supply of cider over the roof and walls.

The origins of this fire have been in doubt, as they so often are in large blazes, as the evidence literally 'goes up in smoke'. One version states that a cow kicked over a lamp and set fire to the straw in a barn. Another story concerns Peg-leg Sullivan. He was said to have gone into Mrs Patrick O'Leary's barn which housed the cow said to have started the fire, but according to this version Peg-leg Sullivan lit his pipe and accidentally ignited the straw. As he tried to escape his peg-leg caught in a crack in the floor; unstrapping the leg he hobbled to safety clinging to the cow.

It appears, as with the fire of London, that the Chicago holocaust started from trivial beginnings. Both seem to be marked by carelessness and it is clear that even in 1871 fire-fighting equipment was poor compared to present-day equivalents.

Fire-fighting technology
Since the burning of Chicago there have been major advances in fire-fighting technology, the most obvious being the introduction of motorised rather than horse-drawn pumps. In addition, the pumps themselves can deliver much larger volumes of water and to greater heights. The standard appliance used in today is able to throw jets up to a height of 150 feet at a rate of approximately 1,000 gallons per minute.

Nevertheless, in spite of all advances, water remains the chief weapon against fire, although in special circumstances other materials such as dry powders and chemicals are used. Currently there is a whole new range of improvements being considered, such as helicopters to fight fires, either in remote country areas or to assist by landing on the roofs of high buildings. The turn-table ladders, although in their day they represented real progress, may be superseded by the 'snorkel', a hydraulic platform like those used by workmen servicing overhead street lighting. This has the advantage of greater manoeuvrability as well as that more fire-fighters can be accommodated on the platform than at the top of a turn-table ladder. High-pressure hoses operating at 600 pounds per square inch, compared to the standard pump which operates at a level of 150 pounds per square inch, have been introduced. These super-jets can be modified so that a mist rather than a solid spray is produced. This is of particular value since the individual water particles are able to absorb a greater amount of heat than the conventional spray with its larger droplets.

The main areas in fire-fighting under study concern communication. In most fire-fighting operations today four appliances are sent, but as most fire stations have only two engines, more than one station must be alerted, and thus communication can be a problem, particularly when the conflagration is large enough to require eight or ten pumps. Obviously a more extensive alarm call has to go out, and the situation can become even more confused when there are several fires in the one area. Thus it is important that as soon as the severity of the blaze has been determined, the appropriate number and types of appliances in the area can be called into action. Any advance which will allow time to be saved in deploying forces appropriately will lead to a decrease in fire losses both of property and lives. Mobile transmitting and receiving equipment are obviously helpful in this connection and the use of automatic apparatus and computers also help to keep track of outbreaks, besides indicating at any particular time just where and how the appliances of different types are deployed. Such a system is in the process of revolutionising fire-fighting.

All these advances, however, have failed to halt the upswing in fire losses. In Britain these are so worrying that the government in 1973 had a booklet prepared called 'Danger from Fire – How to Protect Your Home' which was put through every letter box in the country. In order to draw attention to fires and legislation for fire fighting and fire protection services the Post Office issued in 1974 a set of stamps to commemorate the bicentenary of comprehensive legislation on this matter. It is my view, however, that the rise in maliciously caused fires, which is of the greatest importance, has received insufficient publicity. We shall find out just how serious the situation is throughout the world in the next chapter.

Chapter Three

The Destructive Power of Fire

Since I first began to appreciate the ubiquity of fire I have noticed that on any day of the week in just about every newspaper one is bound to come across at least one item on 'Fire'.

> Accident verdict on six killed in second café fire.
> Jailed . . . the 'hero' of the £25,000 blaze.
> Petrol bomb raid on family.
> *Evening News*, 6 January 1972

On a single day these three headlines appeared; fire had impinged on the lives of three different groups of people. This gives us some idea of the problem of fire, but to get a clearer view we must examine precise figures. Each year in Britain the report of the chief inspector of fire services is published, giving statistics of all aspects of fire and fire-fighting. In England and Wales every year the fire brigades attend to about a quarter of a million fires. In 1970 the total was 244,991, of which 107,397 were of major proportions, 655 members of the public dying as a direct or indirect result of the outbreaks, while 4,100 were injured. It is difficult to give an exact figure for the financial cost of fires but it is estimated that the overall sum, which includes cost of fire insurance and of fire-proofing buildings as well as the direct loss due to the destruction caused by the fires, runs into the region of £500 million. Actual financial loss through fire is probably now in the order of £120 million, the figure for 1970 constituting a steep rise from the 1955 figure which was £50 million. The figures for other countries are also available, published by the National Fire Protection Association of the United States. They give per capita loss trends and show that these are on the increase in every country. The 1966 figure for the United Kingdom was $4.19 per capita, rising to $4.79 in 1970. In the United States the figures for the same years were $9.50 and $12.81. In practically every country there has been a similar sort of rise. Of course it could be argued that this rise merely reflects inflation, but this is not the full story as there has also been a particular increase in the number of larger fires reported over the past ten years. If we examine fire brigade statistics for England and Wales the figure was 62,460 in 1960 and 111,960 in 1969. This increase contains another and highly important aspect: the number of fires which were started deliberately in these two years was 676 in 1960 and 4,102 in 1969. We shall return to these statistics in the next chapter.

Aeroplane fires

Aeroplanes present special fire risks because they contain huge volumes of fuel on board. In the days when piston-driven planes were used, petrol, itself a volatile material, was the fuel, but the jet engine uses a fuel of even higher volatility and there is even more danger not only of a rapidly expanding fire but of explosion. In a study on airport disasters it was noted that about a third of all accidents involving civil transport aircraft occurred within a few miles of an airport, and many on take-off when the fuel load was at its maximum. Further, an analysis of thirteen accidents in which fire broke out after landing showed that only 5 per cent of the passengers were injured fatally as a result of the force of the impact, but over 40 per cent died as a result of the fire. The hazards are twofold. First, the large quantities of fuel carried are in the body of the plane so that the passengers are literally sitting on inflammable material. The second concerns the fact that the passengers are trapped in a closed cabin, so that when fire breaks out poisonous fumes fill the cabin from the smouldering materials used for covering seats, the curtaining, and so on. This problem of course exists also in buildings such as theatres and dance halls, but it is particularly serious in aircraft. Changes in design and the elimination of potentially dangerous types of material have been studied, as have the possibility of crash-resistant fuel tanks and the development of fuels which jellify rather than evaporate when exposed to air. This would prevent the danger not only of fire but also of explosion.

Another problem concerns the means of escape, since in all confined spaces when fire breaks out panic is apt to occur, and unless the crew are well-versed in evacuation, great difficulties can arise. It should be borne in mind that there may be little or no light, the doors may have been jammed in the crash, the cabin staff may have been killed and the stunned passengers may have no clear directions and, in any case, may be unfamiliar with the layout of the inside of the aircraft. It is estimated that approximately 90 seconds is needed to evacuate most modern types of aircraft and this of course involves a high degree of efficiency. Fires start with great rapidity and it may be that in fact only 30 seconds are available in which the passengers can make their escape.

Fires on board ship
As with aeroplanes, so with ships there are particular and special fire hazards for a number of reasons. Many ships carry dangerous cargoes, for example oil, or fertilisers, or highly inflammable materials such as cotton, and this is one reason why fires and explosions are not uncommon on ships. These may sometimes take place on the quayside during loading and unloading, and then the fire can spread on to the land, killing quayside workers and causing damage in and around the harbour area. But the control of fires on board ship also presents its own particular difficulties. In earlier times some of the worst fires happened because ships had wooden hulls and were propelled by steam. One notable example was a conflagration on board the *Amazon* which was totally burned out in 1852.

It may seem that a ship at sea has a great advantage in that it is surrounded by water, but though this is certainly the case nevertheless water cannot be pumped indefinitely into a ship as inevitably this will upset the stability of the vessel, causing it to keel over. For this reason a whole system of international fire regulations has been drawn up so that the design of ships can take into account particular dangers. The main stairways have to be protected with fire-resistant materials, and there must be water-tight bulk-heads which will contain the water in particular areas. Sometimes as a result of a fire the temperature of the hull of the ship rises very steeply and the whole structure becomes red hot. Perhaps surprisingly, nowadays few fires start in the engine room, most having their origin in the passenger and crew areas, bars, dining-room, hairdressing shops and so on. In earlier times voyages were extremely hazardous and many ships and lives were lost, but accidents still occur, as for example in May 1972 when the *Royston Grange* was in collision in the *River Plate* estuary with a Liberian oil tanker, the *Tien Chee*, in dense fog. The *Tien Chee* was loaded with 20,000 tons of crude oil and as a result of the collision oil was thrown on to the *Royston Grange* which ignited on the deck. In no time the whole deck was ablaze, effectively preventing the escape of the passengers and crew aboard, so that all seventy-three perished.

With the developments in shipping, particularly with the building of very large oil tankers, new problems have arisen. Within seventeen days in December 1969, *three* giant super tankers were ripped apart by mysterious explosions. It took three years and a million pounds to find the cause, which turned out to be the fault of the high-pressure water jets used for tank cleaning. Apparently when these jets are played on the side of the tank and strike the metal struts, a thundercloud effect is created and static electricity builds up, the metal struts acting as lightning conductors. The spark then ignites the oil vapour which remains in these tanks even when 'empty'. These explosions are extremely serious, especially when it is considered that the volume of the tank is so great that it could accommodate St Paul's Cathedral with ease.

How do fires kill?
It is a child's early experience that touching a hot stove is painful, and quickly the lesson is learnt. Nevertheless, every year many children are scarred or even die as a result of fire, either by direct contact with unscreened or improperly protected gas or electric fires, by upsetting paraffin heaters, or through their nightdress catching alight while standing in front of an open fire. There is of course legislation which covers the manufacture of domestic appliances and clothing but, in spite of this, over the past decade or so between six hundred and seven hundred people a year have died in Britain alone as a result of fire or the injuries they sustained, and the figures for 1970 show a rise to 839. As in previous years about 75 per cent of these incidents are due to domestic

causes and in 1970 166 children of under fourteen were involved. The elderly are also particularly susceptible.

The destructive power of fire and its magical qualities clearly recommended it to the authorities in medieval times when burning at the stake was a common punishment for both witches and heretics. Bishop Latimer was one such and his words as he died at the stake have been preserved (see page 74). The actual cause of death in those who suffered this terrible ordeal by fire is uncertain. However, being tied upright, as in crucifixion, might lead to fainting, a state which could also result from a sudden acute pain occurring soon after the flames were kindled. Quite quickly with the destruction of body tissues by fire severe disturbance of the circulatory system would occur and death follow rapidly.

Fire can also cause death by asphyxiation, as we shall see, but it has other effects too. If the heat is severe, the skin and the underlying tissues, including muscle and bone, may be destroyed. The main problem then is that the protective sheath around the body is penetrated. Fire has two main effects. First it destroys the nerve endings which, sensitive to pain, are deactivated and therefore the person is no longer able to react and pull away from the dangerous situation; we see this when an elderly person has had a stroke and is unconscious or when a person has an epileptic fit which renders him or her temporarily in the same state. The person cannot respond and an electric fire, for example, if it is in contact with the flesh will continue to burn deeper and deeper. The other effect is loss of the protective power of the skin. This allows infection to enter. The layers below the skin are moist and warm, containing exactly the food materials which the bacteriologists use in the laboratory to grow micro-organisms; hence they quickly multiply and add to the destruction wrought by the fire itself.

If more than three-quarters of the body area is burnt the outlook is hopeless, not only because of the danger of infection but also because the loss of fluid containing protein and vital minerals is so rapid that even its replacement by drip feeding directly into the veins cannot make good the loss. This is a type of 'shock', leading to failure in the efficiency of the heart which is then unable to maintain the supply of oxygen and food substances to the brain and other vital parts of the body. Death quickly ensues.

Means of combating these effects of severe burning are legion. All sorts of ways of covering the surface of the body have been devised, as well as of replacing the lost fluids rapidly and, using special substances to combat infection, employing the hovercraft principle to support the body on a cushion of air so that the surface is not contaminated or damaged by contact or pressure. These techniques are, however, rarely successful when over 75 per cent of the body is involved, and even when smaller areas are damaged the death rate is still high.

Recently the successful treatment of a ship-worker, Riko Roupsa, who had lost 80 per cent of his skin surface was reported. He was placed in a sheet of perforated aluminium foil and warm air was pumped through the holes to form a cocoon in the bed. The air dried the wounds and the same process re-formed the surface layer so that infection could not enter. On the twelfth day after the accident the patient was placed in a bath to clean his wounds, a process which was then repeated daily.

However, even when a person recovers from the initial shock, treatment is difficult because limbs which are not moved become stiff and useless. Exercises therefore have to be started as early as possible to prevent permanent changes in the joints which would lead to virtually immobile arms and legs in spite of the person having survived the initial onslaught of the fire.

Another part of the treatment of Roupsa, as with many others, was plastic surgery. This involves the 'transplanting' of skin from the unburned areas of the body or from parts where the burns are slight and the skin has re-formed. In the case of this particular patient, the damage had been deep and therefore healing was slow and repeated operations were necessary. Roupsa was lucky, he survived his burns, but still there are after-effects – stiff limbs and withered muscles which require prolonged treatment.

There are other ways in which fire can be lethal, apart from burning. Some victims caught in a hotel or factory building may, after the fire has been extinguished, be found dead but without any sign of burns on their body at all. Why have they died? Their blood will be found to contain a high level of carbon monoxide, a poisonous gas generated by fires in confined spaces; there is insufficient oxygen to burn woodwork and furnishings and as a result this lethal gas builds up.

There are other problems as well. A tragedy caught the headlines in 1970 when in St. Laurent Dupont in France 145 teenagers were killed when a dance hall caught fire. A lighted match kindled a plastic chair-covering and the blaze quickly spread. The plastic decorations, when subjected to the high temperature, gave off a noxious vapour and many of the teenagers were asphyxiated by the fumes. The problem of plastic materials was also important in the fire on the Isle of Man in 1973 and this, as we have mentioned, is the type of thing which happens in aeroplane cabins when a fire begins. Oddly enough, even materials which have been fire-proofed are likely to give off these poisonous vapours.

The forensic expert

The forensic expert is called in to establish the cause of death, whether from fire or by other means. If a body is found at the scene of a fire, in a house, motor car or factory building, the cause of death has to be determined. It may seem straightforward, and be in fact directly due to the fire, but this is not always the case; the person may well, for instance, have been crushed by falling masonry. It is necessary also to determine whether the injuries were sustained while the person was attempting to escape from the inferno or perhaps while he was trying to rescue someone else.

The problem is confused by the fact that when burned the skin contracts, especially at the points where clothing is tight. When clothing is removed at the *post mortem* examination it appears as though there has been a tight band, for example round the neck, where the collar and tie have been, which might suggest that the victim had been strangled. Further important questions are then posed. Was the body placed in the fire after death or had the victim been violently attacked and left unconscious and then the house or car set on fire? This, like so many forensic problems, needs careful assembling of all the evidence, but there are laboratory tests which can also be of value. The carbon monoxide level in the blood is one. If there is 50 per cent or more (or according to some authorities only 5 per cent) then the person was alive when the fire started (see Gradwohl's *Legal Medicine*). Another important point is the presence of soot in the air passages when the *post mortem* examination is conducted. This, too, would indicate that the person was alive when the fire started as he had inhaled smoke particles.

These and other questions about the dangers of fire, as well as its forensic aspects, will be examined in the next chapter.

Chapter Four
Fire-Raising and its Investigation

The fire investigator
Someone has to try to sort out what is the cause of a particular fire. This is the fire investigator, who is responsible for locating the faulty wiring or for finding the cigarette end which charred the upholstery. His solution involves knowledge of such diverse subjects as architecture and chemistry, whether the building was fire-proofed, for example, and the materials used in fire-proofing. But this is not all; he must also find out if there were any underlying motives suggesting deliberate fire-setting. As Kirk says in *Fire Investigation*, there are people who 'wish to vent their perverted feelings on humanity, wish to obtain insurance money, wish to destroy property and people'. Sometimes detection of these deliberate fires is not easy, and to learn something about this subject we must turn to the experts for their comments.

E. B. Chambers of the Fire Research Station has made a study of incendiarism; studying the fire statistics as reported by fire brigades in 1964 he found that 87.2 per cent of fires in buildings were reported as accidental, 1.3 per cent were of unknown cause, and only 1.5 per cent were attributed to malicious incendiarism. Not dissimilar percentages were noted for fires in the open. However, when the figures relating to financial loss were studied, it was found that maliciously started fires caused much more damage than did fires of accidental origin. The reason for this is quite clear. If a plumber's blow lamp sets fire to the lagging round a pipe, then if the workman cannot immediately put it out himself he will promptly call the fire brigade. On the other hand, if an arsonist sets fire to a wastepaper basket in some well-stocked showroom he will be the last to do so. Another problem is that fire investigation has tended to concentrate on the means of ignition of the fire, rather than the possibility of a malicious igniter. Hence the official figures tend to give an underestimate of malicious fire-setting.

The Menai Straits Bridge fire
Stephenson's tubular bridge over the Menai Straits which links the Isle of Anglesey to the mainland is a vital route for traffic between Britain and Ireland and the only rail connection. This bridge caught fire on the night of 23 May 1970 and was so badly damaged that it had to be re-designed before building could commence; it was not reopened until 30 January 1972. It was a spectacular fire, but how did it start? Was it accidental, a spark from a passing engine? No, steam engines had been withdrawn from service years before. Was it part of a childish game or the act of a deliberate fire-raiser? To determine the answers to this question it is necessary to see where the fire started and how it spread.

The main structure of the original bridge was of wrought-iron in the form of tubes. The upper part of the bridge was made of large quantities of timber which, in order to protect it from the weather, was heavily tarred. In the hundred and thirty years or so that the bridge was in operation fires had not occurred, in spite of the fact that steam engines had regularly showered the bridge with sparks, because when steam engines were in operation the railway had a look-out on the bridge to watch for fires. However, this practice was abandoned when steam trains were withdrawn. As a result the fire in May 1970 was well alight before anyone knew and, in any case, it presented great difficulties for the fire-fighters because the only water was in the straits far below. So in spite of the efforts of the brigade, it soon became obvious that the fire would have to be allowed to burn itself out. The bridge became red hot in the process, warping, creaking and groaning, but in fact it did not fall.

Before the fire was really established eye-witnesses saw two suspicious events: a man leaving the mouth of the tunnel, apparently with a gun, and later a group of five youths. The man with the gun was traced. It was in fact a toy pistol he was carrying, but the most surprising thing about him was his past record, which was one of malicious fire-setting. Apparently on this occasion he was not responsible. The culprits were the five boys, who had struck matches and set alight to paper lying around and then run away allowing the fire to establish itself in the rubbish that lay on the railway line. The flames ignited the wooden roof and the conflagration rapidly spread. The reconstruction of the bridge cost millions of pounds. This indicates how a boyish prank can turn into a costly disaster. The Menai Straits Bridge can thus be said to have been destroyed by an accidental fire, in that it was not intended by the boys that the bridge be destroyed, but we must look further into the statistics to see just how many fires are in fact caused deliberately.

Fires in letter boxes
The case of the Menai Bridge fire was solved, but sometimes the source of the fire is never found and it is pure supposition as to whether it was accidental or malicious. Nevertheless, it is usual to assign a cause in the official statistics which are published annually. How inaccurate these statistics are can be seen in a further investigation by Chambers of 'supposed causes of fires in GPO letter boxes'. He points out that it is inconceivable that fires in letter boxes can be of an accidental kind. In spite of this only four of 184 such fires in a particular year were reported as due to malicious ignition and most were attributed to pranks of children. The actual causes listed were forty-nine explosives and fireworks, thirty-two matches due to children, twenty-four tapers, lighted paper and sticks, twenty-three matches, eighteen smoking materials, the rest miscellaneous and unknown causes. It seems highly likely that more than four of these fires were of malicious origin, but as we have already said, more importance is attached to finding out the means of ignition than the motive of the igniter.

Another way of discovering the under-reporting of malicious fire-raising is to find out from people actually convicted of arson about previous episodes in which they have been involved. One such study revealed that thirty-three fires had been deliberately started, only seven of which had been included in the published statistics as fires due to the work of a deliberate fire-setter.

Some years ago a detailed analysis of the main causes of fires was undertaken in the West Riding of Yorkshire. In the course of the investigations another interesting piece of evidence on under-reporting of malicious fire-setting emerged. Two series of fires were discovered which had spanned many years. Fry reported them as follows:

> The first series was set by a man aged about thirty-five. There were twenty-nine fires in all and they covered the period from September 1955 to February 1960 with a gap of two and a half years in the middle when it turned out that the man had been imprisoned for some other offence. As soon as he came out of prison the fires began again. However in the official statistics twelve of these fires had been recorded as being due to accidental causes, nine were said to be of unknown cause, five didn't seem to be recorded at all and it was only the three at the very end of the series which were noted as being intentional. Sometimes the malicious incendiarist has a particular predilection for a certain type of building or situation. However in this particular series farms, mills, a church, a school and a variety of wooden huts were all chosen. The only point was that buildings without any kind of occupant were chosen.
>
> The second series was smaller. There were nine fires but the same sort of thing happened again. Three were recorded as accidental, two of unknown origin and the last four of intentional origin. The period covered was from 1962 to 1963. The series of fires was set by a man of about twenty and again he set fires in places that were convenient rather than choosing a particular type of building. Timber huts, joinery works, mills and places of that kind.

Losses from maliciously set fires

The problem is compounded by the high financial loss of maliciously set fires. This emerges clearly in the investigations carried out by the Fire Research Station in conjunction with the Fire Protection Association. This voluntary organisation is one which gives advice on fire protection and other matters and is supported by the insurance companies who are concerned with reducing fire losses as well as investigating the causes of the fires. All fires in which the losses amount to £10,000 or over are investigated. In the year 1970 there were 1,115 such fires, total losses from which reached £69,944,000. The three main causes of these fires were faulty electrical wiring (139), malicious intent (124) and carelessness with cigarettes and smoking materials (65). Though the number of fires started intentionally was smaller than that due to electrical faults, the scale of damage was approximately a third more. Not only that, but the number of malicious fire-setting incidents has gone up by six or seven times in a decade. The *Fire Journal*, published by the National Fire Protection Association of the United States, also produces some disquieting figures. They note that 'the number of fires per thousand population has increased only slightly over the past ten or more years, while incendiary and child-related fires and those of unknown causes have increased. They have been offsetting decreases in fires from smoking and from unsafe heating and cooking equipment.'

Who determines the cause of fire?
In the first instance, the fire brigade experts report on the cause of the fire, and of course, in most instances, an insurance company is also interested. If there is any doubt the expert is called in. The fire investigator, with his detailed knowledge of buildings and the response of particular structures to fire, is summoned. He must study the site after the fire to determine, for example, whether there were multiple places of origin for the fire. Such a feature will generally be against a 'natural cause' such as faulty electrical wiring. Interrogation of witnesses is also essential and often gives a vital clue to malicious fire-setting for it helps to determine where the conflagration started and how it spread. Examination of any body or bodies found in the building can provide the forensic pathologist with evidence which may in turn be useful to the fire investigator.

Sometimes the malicious fire-setter may be found dead at the site of his activity. If highly inflammable liquid, such as petrol, is used then the flames may spread with great rapidity, and rather than a straightforward blaze, explosions may occur, in which the arsonist perishes. The fire investigator often has a difficult problem on his hands in determining the cause of the trouble, since the flames, heat and smoke may destroy almost all the evidence. It is quite likely that fires of deliberate origin are in fact reported as 'cause unknown' since evidence to support malicious intent is not forthcoming.

The act of deliberate fire-setting is known as 'arson'. It is defined in Chambers's *Twentieth Century Dictionary* as 'the crime of feloniously burning houses, haystacks, ships, forests or similar property'. Arson is a legal term and covers the whole range of deliberate fire-setting, but there is no single equivalent word that the psychiatrist can use, since the motives or apparent motives are so varied. In some instances they are quite clearcut, for instance to cover up crime or to collect insurance money; again, some malicious fire-setting has political motives. But in other instances there are no overt motives and the act of fire-setting is apparently an end in itself. But in the first instance let us consider the problem from the point of view of the fire investigators.

The arsonists
The arsonist obviously has to enter the premises which he wishes to set on fire. On the one hand it may be that he is an employee who is already in the building or who at the week-end gets in with a key he has obtained fraudulently. On the other hand the fire-setter may have to enter illegally and then there will be signs of how he has achieved this, for example a broken window, and then fingerprints may be a vital clue for the investigator.

Once inside, the problems are not over, for like the man who lights a fire in his own sitting-room the arsonist wants the fire to become well established but at the same time, at least in the first instance, controlled. If the speed and spread of the flames is too rapid he himself will be trapped and unable to escape. The basic requirements for the propagation of a fire are fuel and ventilation. The other elements needed are of course the means of ignition, and a way of maintaining the flames until they spread to the fabric of the building itself. The fire-setter may simply use matches, or a cigarette lighter, douse a full wastepaper basket with petrol and leave it at that. However, delayed methods are sometimes employed, for example candles or waxed tapers which burn down slowly to the inflammable material which is to form the nub of the conflagration. Ingenious methods of ignition, chemical and electrical, have been tried in the past, but of course they may not be reliable and they may give valuable clues to the investigator since the initial source of the fire may in fact on occasions be relatively unscathed in the final holocaust. Most often some type of liquid fuel is used to ensure that the fire gets a hold quickly. The fuel used depends partly on the arsonist but also partly on availability; for example industrial alcohol and turpentine can easily be obtained in factories. Duplicating fluid has been employed in starting office fires, but paraffin is on the whole less favoured than petrol. It is surprising, however, that these fluids may not themselves be entirely consumed, even if the fire is a large one, and the investigator may be able to identify them days or weeks later. The liquid soaks into the floor boards, the pile of rubbish or even into the soil if the fire is started outside.

The choice of a site within a building to kindle a fire can be difficult, since there are strict regulations about fire prevention and walls have to be coated with fire-resistant paint, and ceilings must be made of non-inflammable materials. Basements are the places selected most often by arsonists as they allow easy access and egress. Also there is usually good ventilation along a basement corridor.

The arsonist who has had previous experience will in all probability be aware of fire-fighting and fire investigation, and is therefore likely to start his blaze near a radiator or a heater, or in a grate so that casual inspection may suggest that the fire was accidental. Some are not content with one fire, but start several, and this can be a most important clue to the investigator that the outbreak was deliberately started.

The fire-raisers can be divided into four main groups, and each will be described in more detail in the chapters that follow. The first group have a reasonable explanation for their fire-setting activities, while the other groups show progressively less convincing motivation.

The first group includes those in whom the motive for fire-raising is quite open: the people who use fire to cover up other crimes or which lead to financial reward. In the 1930s there was the famous case of Leopold Harris. This will be examined, and similar cases of more recent times will also be looked into.

My second group of fire-raisers seem also to have good reason for their nefarious and destructive activities. These are the political fire-raisers. The burning of the Reichstag is perhaps the prime example, but there are many other ones. Northern Ireland at the present time furnishes clearcut examples of political fire-raising; here the damage runs into millions of pounds each year.

An overall element of desperation can be sensed in this second group and this is even more apparent in the third group of people, those who kill themselves by burning – self-immolation. This strange and horrible method of suicide has many facets. The death of Jan Palach in Czechoslovakia, for example, shows a largely political motivation. One's first reaction might be to regard all suicides by burning as occurring only in the severely mentally deranged, but Jan Palach at least was not one such. But there is another aspect of suicide by fire; it could be regarded as an act of submission to the flames, the ultimate masochistic act.

The motives of the fourth group can only be called deeply perverted, however buried these motives may be. Fire-setting for 'kicks' is one reason given by people in this group, but there are others in this group, too, such as the person who feels aggrieved with his employer and burns down the factory, or in the past the country lass who had fallen out with the squire and set his hayricks ablaze. There are also those who, in the depths of a depression, burn down their houses or during some severe schizophrenic illness start a fire at the instigation of a hallucinatory voice. Perhaps the most difficult of all to understand is the 'firebug', whose compulsions lead him to start a series of fires, sometimes dangerous to life.

Human behaviour is inevitably complex, and that of deliberate fire-setters seems to be particularly so. Nevertheless, it is important, in the face of the increasing number and cost of maliciously started conflagrations, to attempt to understand such people, and to some extent classification of motives may make it possible for us to come to some understanding of these tortured individuals who, often through some fault in upbringing or unfortunate experience, are led to bizarre, destructive and often life-threatening behaviour. Perhaps they can even be helped.

PART TWO

THE MOTIVATED FIRE-SETTERS

Chapter Five
The Profit Motive

The Leopold Harris gang
A taper slowly burnt down towards two flat photographic trays. Suddenly there was a mass of flame, ignominiously doused by a bucket of water. This was the beginning of a profitable adventure which was to last for years and end in the Old Bailey. It was November 1927 and the place, 196 Dean Street, Manchester, the premises of Fabriques de Soieries Ltd.

The whole building was dark and silent apart from the boardroom where two men and a woman had been watching the taper intently; an ingenious method of fire-raising had been devised by them through which they hoped to make themselves a lot of money. The blaze occurred after a delay of fifteen minutes, the time taken for the taper to burn down from its tip to the trays containing the inflammable photographic chemicals – a valuable quarter of an hour in which to escape from the neighbourhood and allow time to build up an alibi. The trio then moved into the stockroom. While the lady stayed inside the men checked from the street that the light from the candle she held was invisible through the shutters. That was the final touch; they were now ready to carry out their incendiary activities and only the date had to be decided. This whole incredible saga of fire-raising intrigue is told by Harold Dearden in *The Fire Raisers – the Story of Leopold Harris and his Gang.*

Who were this trio and what were they about? One of the two men was middle-aged, and went by the name of Louis Jarvis, alias Jacobs; the younger man was an Italian, Mr Camillo Capsoni. Both were directors of a silk business. Mrs Bing was the secretary and a friend of the younger man.

Mr Capsoni had been involved before in companies of doubtful repute and was keenly seeking some kind of lucrative reward for what he thought of as his talent. Some years previously he had met Mr Lewis Jarvis, with whom he had done business, and had heard of a fire at his premises. He naturally expressed his sympathy but was surprised when Mr Jarvis commented that 'there are fires and fires'. Mr Capsoni was immediately interested and thus the plot at the silk business was hatched. Mr Jarvis saw a successful commercial organisation as one with 'inexpensive stock and a good fat fire policy'. Of course, burning the stock and the building which housed them did present difficulties which Mr. Capsoni appreciated, but it was the introduction of 'Mr Leopold Harris' that dispelled any doubts he might have had about the enterprise.

Mr Harris, according to Dearden,

> was what is known as a fire assessor and Mr Capsoni gathered that his life was a very busy one. When people had fires Mr Harris was in the habit of appearing on the scene of the disaster with quite remarkable celerity, offering to secure for the victim the maximum amount of compensation from any company with which the latter might happen to be insured. He was paid for these services on a percentage of the sum received by the victim. He had a large business which had been established by his father and he had an even larger and more valuable circle of influential friends. Certainly, quite apart from his charm of manner his popularity was well deserved.

Having elicited the help of Mr Harris, Jarvis and Capsoni stocked their showroom with *crèpe de Chine* which they picked up cheaply in Leon because it had recently emerged from someone else's fire! The premises' merchandise, tastefully arranged, was photographed and a print sent to Mr Harris.

7 November was the date fixed for the event. The trays and taper were set up in the stockroom and lit. It was about 6.15 when the 'employees' we have already met left in a leisurely fashion. Mr Capsoni strolled to the Midland Hotel where Mrs Bing was waiting for him and where Mr Harris had conveniently put up. He was always early at the scene of any disaster and this eventually, although many years later, was his undoing.

Unfortunately, when the fire started the fire brigade also acted with alacrity and the blaze at Fabriques de Soieries turned out to be a rather small one. Nevertheless large volumes of water were used and the stock suffered considerable damage. Mr Harris assessed the damage at £32,000, and a claim was forwarded to the silk firm's insurance company, while Mrs Bing and Messrs Jarvis and Capsoni set to work concocting the invoices to substantiate the claim. These however were not needed, and the matter was settled at £29,000, of which Mr Jarvis got £12,000, Mr Harris £8,000 and Mr Capsoni a mere £1,000.

But truth will out – usually. Mr Jarvis had the habit of making entries in a book called 'The Confirmations Book', tastefully bound in leather and showing samples of the material in the warehouse with the amount and price for which it was obtained. All other records of the silk wholesale business were destroyed but this special book remained in Mr Jarvis's possession and was finally used in evidence by the crown, at the Old Bailey, about five years later when Mr Jarvis was sentenced to three years' imprisonment.

It is important to bear in mind that though Mr Harris was involved in dishonest work based on his fire-setting activities, his income was largely derived from his lucrative fire-assessing business, which had been established by his father and was perfectly reputable. Mr Harris was always regarded as scrupulous and punctilious in his dealings with clients; it was only when he came across one whom he judged to be less than honest that fraudulent claims arose. He was, in Dearden's view, 'a singularly shrewd judge of human nature; he could sum a client's honesty almost at a glance; and only when he knew himself to be on safe ground did he suggest anything in the nature of a fraudulent claim and increase his fee by taking a share of the booty extracted by these means from the insurance company.' One of the problems with Mr Harris and his work was that it was uncomfortable. Fires often happened at night and he had to leap from his bed and rush to the scene of the disaster. As he was a home-loving and highly civilised man he set about modifying his inconvenient occupation by arranging the times and places when the conflagrations occurred!

There was another side to Mr Harris's activities. His sister was married to a Mr Harry Gould who had an interesting part to play in the fire-setting business. Mr Gould supplied, say, £1,000 worth of junk, and invoiced it to the firm which was to be destroyed at ten times that value; after the fire Mr Gould bought back the goods at a reduced price because they were 'shop-soiled', often re-selling them to be 'burnt' again!

Mr Harris's next fire was to be at the 'Continental Showrooms' at Leeds at 12 Bassinghall Street. This was a business set up with Mrs Bing in charge especially with incineration in mind. It may seem odd that this group could continue with their activities without being caught. But it must be remembered that their fires were examined not by a single person but by a whole series of different people, each working on behalf of a different fire insurance company acting independently, to a large extent, of its fellows. A shrewd observer would almost certainly, at this stage, have noticed the similarities in all the fires; and indeed this was what eventually happened.

An even more extraordinary fact that escaped notice for a long time was an early attempt which Mr Harris perpetrated to defraud an insurance company. It preceded his fire-raising activities, and unlike them proved unsuccessful. He obviously saw his clients getting rich from his careful work of assessing so he decided that he too would attempt something of the kind. He staged a burglary at his home and claimed a modest £1,500. Strangely enough, the insurance company refused to pay and Mr Justice Branson who presided said that this was clearly a bogus claim and the case was dismissed. It is odd that Mr Harris who already worked in the insurance business should not instantly have been put out of business, but he wasn't; in fact, his business expanded.

The 'Continental Showrooms' was stocked following a tour in Italy and France by Mr Capsoni and the date was then set for the fire. On the particular day chosen Mr Harris had an engagement, so it was postponed till the next day. On the other hand Mr Capsoni was not so lucky for he had to cancel an appointment that he had made long before, to be available. As Dearden points out, this unimportant detail 'illustrates the spirit of ungrudging loyalty and service which prevailed in Mr Harris's unique organisation'. Unfortunately the fire was not to be a great success. At least, the incineration was, but the insurance claim was not, a mere £2,000 or so being finally obtained. It was a bad year, 1929, for Leopold Harris.

The next conflagration at Alfred Olton Ltd, was well planned. The premises were stuffed with rows of fur coats, most of them, in the words of Mr Smith who worked in the company, 'all rubbish'. There was some difficulty in disposing of Mr Smith before the fire was started – he was sent away on holiday – and then a claim for £9,000 was placed with the insurance assessor, who had already been bribed with a good-quality fur coat. The whole affair was wound up satisfactorily by the sale to Mr Simon Wolfe of the goods on the basis of 75p in the pound.

Now it was Mr Capsoni's turn to set up the 'Franco-Italian Silk Company', assisted by Mr Harris and Mr Harry Gould. Stocks of silk, damaged by water and badly scorched, were brought in to cram the storerooms, and an excessive fire policy for £10,000 was taken out. It was at this stage that Mr Capsoni married Mrs Bing, who was offered the privilege of acting as the incinerist of the Franco-Italian Silk Company.

Mrs Capsoni was entirely successful, though there were a few difficult moments when the independent assessor, Mr Loughborough Ball, quizzed Mr Capsoni on the likely cause of the fire. Nevertheless he agreed that it was accidental and arranged for a cheque for over £21,000. Mr Harris took £10,000 and the rest of the spoils were divided among the others.

Mr Harris had many friends who were necessary for his activities, because although Mr Harris was called an insurance assessor in actual fact he was a claim-maker on behalf of the client. The insurance companies themselves put up their own assessor who had to make certain that the claim was justified. He also had to satisfy himself on the original value of the property and the goods as well as ascertaining that the fire was accidental. Mr Harris clearly needed, in his 'business', to have friends in high places.

It was therefore a stroke of luck for him when he happened to meet a certain Captain Brynmore Eric Miles, at the Old Bailey, when he dropped in to listen to a trial in which an insurance assessor had refused a bribe and the two men who had proffered it to him were being tried. Mr Harris exchanged a few words with Captain Miles, chief officer at the London Salvage

Corps, and this proved to be the beginning of a valuable if anxiety-provoking friendship.

To understand the importance of this chance encounter it is necessary to look for a moment at the fire-fighting organisation as it then was and the role of the London Salvage Corps. We have already seen how, following the Fire of London, various fire offices were set up with their own fire-fighting services, and how the fire marks attached to buildings indicated which fire brigade was to be called if trouble broke out. However, this system had soon been seen to be quite haphazard and later a 'corps which should consist of a body of men to attend to fires within the area of the metropolis' was set up. It was Captain Miles who was, in the 1930s, at the head of this corps.

There followed a series of meetings between Miles and Harris, at which it was suggested that Miles let Harris know as quickly as possible about outbreaks of fire in return for money. Apparently, though Miles was provided with free accommodation, lighting and other facilities, he was in serious debt and would appreciate the extra money. The fact that the Salvage Corps had news of every outbreak of fire within a second or two of the call being received by the individual fire brigade, and the fact that this information was to be transmitted to Harris, clearly put him in a very favourable position. His job of making claims was obviously a profitable one, but there was great rivalry between him and the other claim-makers, who were in strict competition; with this new development he would certainly have the edge on them.

There was another side to Captain Miles's suggestions, namely that at times fires had a rather obscure and doubtful origin and the insurance companies might make trouble over claims if they knew these facts. Mr Harris, it was suggested, would be on the scene with alacrity and with his expertise he would be able to remove or conceal any feature which might lead to difficulties! The men came to a 'gentleman's agreement' and Mr Harris congratulated himself on his good fortune.

Captain Miles and Mr Harris met frequently over lunch, and these meetings even came to the notice of the committee who controlled the Salvage Corps. However, Captain Miles extricated himself somehow and Mr Harris was able to take full advantage of the information he received. He profited as a result, but his fellow claim-assessors soon noticed the speed with which he reached a fire and were quick to suspect that he was gaining inside information.

Harris was a generous and gregarious fellow. Money came to him with ease but it also departed in the same way. His many friends were now demanding higher and higher 'fees' but clearly he could not allow a loose tongue in his business and he paid up. There was another difficulty; by 1930 his legitimate enterprise was on the decline. It seemed that his name at the foot of a claim 'was to the insurance companies like a red rag to a bull. Books and invoices were scrutinised in the most unfriendly way imaginable, and since the essence of insurance is mutual trust, it was the sort of thing which must have distressed everyone.' As a result, even genuine claims met with this treatment.

That was not all; another figure appeared on the scene in the shape of Mr William Charles Crocker, a solicitor specialising in insurance work. Mr Crocker disbelieved everything presented to him by Harris, even the most elementary calculations on the simplest of invoices. It was he particularly who made Harris's life difficult and who eventually was responsible for the exposure of the whole racket.

But all this was in the future. In the meantime, Mr Harris was beginning to collect around him an even larger group of incendiarists or would-be incendiarists. There was Mr Herivel, for example, who was sixty-nine and was said to have set fire to two premises in Staining Lane. These were excellent for arson as at both back and front there were narrow streets ideal for obstructing the entrance of fire brigades. Mr Herivel occupied the basement and Mr Harris found a tenant for the second and third floors, Mr Harry Christopher Priest. Mr Priest was someone who was a bit unusual even in the Harris fraternity, for he started fires for pleasure as well as profit. He was a man who could be considered either of 'considerable versatility' or a misfit, depending on one's viewpoint. He had been a marine, a tram conductor and, not surprisingly as we shall find out later, a fireman. Mr Priest had shown some ability in burning down premises but he had also made mistakes in the fire-setters' business. At the scene of one fire a parcel wrapped in newspaper was found which contained scorched articles. This set the inquisitive assessor thinking and the claim was knocked down from £7,500 to £2,700. The goods had been supplied to Mr Priest by Mr Harris's relative and accomplice, a Mr Gould. They were already fire-soiled and had come from a blaze which had already been 'most capably assessed by Mr Ball'.

In early February 1931 Priest set up as a mackintosh manufacturer. The premises were stocked by Mr Gould and they burnt well. The claim was successful, but poor Mr Herivel in the basement had his property damaged by water and felt himself to be most unfortunate to have had so many fires in his premises in such a short time. However, such a successful outcome cheered Mr Harris, who was at a low ebb, and he called some of his friends together for dinner in a restaurant early in April. Priest arrived first and shortly afterwards Mr Capsoni. Harris introduced them and they shook hands.

'Show Mr Capsoni your card, Harry,' said Mr Harris.

Priest fumbled in his pocket and produced a picture postcard of himself in full fireman's uniform. The suggestion (by Harris) that the two should go into business together got the party off to a 'blazing start'.

Mr Capsoni returned from holiday in Italy and after a slim resolve to 'go straight' went into partnership in Poland Street with a Mr Bergolz in a bric-à-brac business. Quite soon he was planning a fire, with Priest keen to take part, but Capsoni (shrewdly, it turned out) decided against this particular partnership. This time a slower method was required than the taper and photographic trays used previously, and on this occasion lengths of film were stuffed into a wastepaper basket, on the top of which a celluloid buckle and a candle were precariously perched. The length of time necessary to reach the film was worked out in an experiment by Capsoni. The fire itself was a roaring success, but the claim drawn up by Mr Harris for £5,863 was rejected completely by the insurance company. Why? Well, Mr Crocker was involved. How did he know that the holocaust was not accidental but had been started by an 'interested party'? He used a simple expedient: he entertained Priest, who when he was filled up with alcohol chattered endlessly. He talked so much that he gave the game away, not only Capsoni's Poland Street blaze but much of the Harris organisation as well: he was always talking of 'the Prince' – a title that suited Harris's grandiose personality. By 1932 Mr Crocker had already discovered that Mr Priest's 'prince' was in fact Mr Leopold Harris. Further, he had observed that Harris's name was associated with very many claims about which Mr Crocker instinctively felt there were suspicious circumstances. But hard facts were difficult to obtain.

Mr Priest, then, was talkative, especially when with companions in his favourite public house in Highbury. On one of these drinking sessions he suggested to a Mr Cornock and a Mr Matthews that they might like to be involved in a spot of fire-raising. They were, it seemed, rather incredulous at first but the boasting Priest assured them that there was to be a fire in Poland Street. Unfortunately, Matthews and Cornock were informers for Mr Crocker, so that when the conflagration took place the claim put in by Mr Harris was known to be for a 'fire' that had been started deliberately. As a result of this information Mr Crocker persuaded two of the leading insurance men, one in Lloyd's and one in the Sun Office, to place money at his disposal so that he could scrutinise Mr Harris's claims as they came in and check the files of past claims. This was possible because the ingenious Mr Crocker had carefully photographed in his office the ledgers and invoices supporting the claims before returning them. At this stage Mr Harris did not have an inkling that the net was closing around him.

Mr Capsoni, one of the earliest members of the Harris gang, had been in eclipse for some time partly because of the legal action he had taken against Mr Lewis Jarvis. This action, which was finally settled out of court, could clearly have led to the exposure of the whole gang, since money had been paid by Capsoni to Jarvis in connection with the Franco-Italian Silk Company fire in 1930. For this and other reasons Capsoni was out of favour and living with his wife in rather straitened circumstances. Then, by chance, they noticed in an illustrated paper a picture of Mr Jarvis; he was fighting the flames of a fire which had broken out in some premises in Wembley. This was the chance for revenge, but Capsoni decided that he himself would stay in the background and send his wife to discuss the matter.

She went to the office of the Scottish Union and National Insurance Company who were responsible for the insurance on the building so heroically defended by Mr Jarvis. There she saw Mr Henderson, the assistant secretary of the company.

Mrs Capsoni informed the assistant secretary that she knew for certain that Mr Jarvis had committed arson on a previous occasion and even confessed that she and her husband had joined Mr Jarvis in such an escapade some years before. Naturally Mr Henderson was interested in the story, but he did not feel that it necessarily followed that the recent fire had been started deliberately. In the circumstances he took the quite reasonable step of arranging a meeting with Captain Miles, the chief of the Salvage Corps. This was the beginning of a drama in which Miles played the central role. He was still in the pay of Harris and as can be imagined had to handle the interview with Mrs Capsoni, and many other related interviews, with some delicacy. One wonders how he managed to live on this particular knife edge for the next few months, with Mr Crocker on the one hand and Mr Harris on the other.

Captain Miles was forced to the conclusion that the only way to cope with Mrs Capsoni's information was to refer her to Mr Crocker. She unburdened herself to Mr Crocker, who was a good listener. Mrs Capsoni told the whole story, not forgetting the Fabriques de Soieries fire and all that led from it. This was in spite of being warned by Mr Crocker that she was incriminating both herself and her husband. Later Capsoni himself also gave evidence to Mr Crocker and so another piece of the Harris jigsaw fell into place.

Mr Crocker had arranged not only for documents to be photographed and carefully studied but also for all the main characters in the drama to be shadowed, with the one exception of Harris who went about apparently unaware that anything was wrong. But all the time Mr Crocker was patiently collating his dossier. Because of the complexity of the evidence, he set it out in the form of an ingenious diagram; on the edges of a rectangle each fire was represented by a circle, starting in the top left-hand corner – fire 1 – and ending at fire 29. In the middle were four circles representing the Harris organisations – the Harris Firm, Harris Relations, Harry Gould and the Z Gang (the exact significance of which is still uncertain). He then drew lines connecting the various features which the fires had in common, the tangled mass of lines shown at the trial becoming known as 'Willesden Junction'. The diagram showed quite clearly that all the fires were related to Harris, either because he had assessed the fires, because he had financed the company involved to stock up, or because his relatives or associates had a connection with the particular firm. The effect of this diagram, which took years of patient research to build up, was to connect apparently unrelated incidents which, because they had been insured with different companies, had not previously been examined together.

By the early part of February 1933 Mr Crocker had completed his task, and he arranged for a series of warrants for the arrest of Harris and fifteen other 'gentlemen'. To complete the coup they had to be arrested as near simultaneously as possible, a task entrusted to a certain Chief Inspector Yandell, who had been tutored on the Harris gang and their activities, so that at the time of their arrest as much evidence as possible could be collected from their homes to support the information which Mr Crocker had so assiduously gathered together over the years. The arrests were completed, with the exception of a Mr Satterthwaite who committed suicide when he got wind of his impending arrest.

The Old Bailey trial was one of the longest criminal trials on record, thirty-three days in all; at the end of it Mr Leopold Harris was awarded fourteen years, the lion's share of the sentences, which totalled fifty-five years. The other defendants were David Harris, Harry Gould, Lewis Jarvis, Bernard Bowman, Felix Bergolz, Harry Priest, William Herivel, James Robert Cross, Leonard Riley, Victor Cope, Bernard Marks, Walter Westward and Adam Ball. Some months later Captain Miles also came to trial and was sentenced to four years' penal servitude.

Are there still fires for profit?
The Harris case involved two main illicit activities which overlapped, the first concerned with the fraudulent inflation of fire claims and the second with deliberate fire-setting, which was of course followed by equally inflated insurance claims. This type of activity had its counterpart in the United States, where there are two main types of illicit fire-raising, the 'arson ring' characteristic of the 'boom' years and the fraudulently inflated insur-

ance claims which occur in the 'years of depression'. The 'arson ring' is one example of the complicated organisation of crime in the United States and concerns two main characters, the 'brains' and the 'torch'. It also involves crooked insurance adjusters and there is an element of protection racketeering in that businessmen are told that their premises will be destroyed by fire if they fail to pay the appropriate sum. There was a famous arson ring in Brooklyn, New York which involved 'Sammy the Torch'. He, like a hired murderer, was completely devoid of normal feelings for life and property, acting merely as a paid agent and collecting his fee when the job was done.

As recently as 1972, Tom Driberg, the Labour MP for Barking, asked a question in Parliament about alleged fraudulent claims for fires in which innocent people had died. He pointed out that there had been an increase in the number of fires in London caused by arson and in some of these, notably the hotel fires, there had been loss of life. Certainly, fire-setting for profit still goes on and it is clear that only by constant vigilance on the part of the insurance company, the fire brigades and the fire investigators, can the public be protected from the serious danger of malicious fires, a form of pollution which characterises the complex, intensely competitive and crowded society in which we live.

Chapter Six
Political Fire-Raising

Burning and bombing are two of the means used by the more anarchistic political factions. Knowledge of the use of gunpowder and other chemicals to make explosions or even Molotov cocktails is not in the hands of everyone, so someone with experience of these must be drawn into underground political movements in order to perpetrate such acts of terror. This is just the way it came about with the infamous Guy Fawkes – who was by no means the main conspirator in the Gunpowder Plot, so that it is remarkable that 5 November is celebrated as Guy Fawkes Day as though he were the arch-villain. In fact, the event commemorates the failure of a group of English Catholics to assassinate King James I more than 350 years ago.

The gunpowder plot
Guy Fawkes was born into a family of ecclesiastical lawyers who served the Consistory and Exchequer Courts of the Archbishop of York; he was therefore among the petty gentry. His father died when Guy was only nine years old. He was, of course, brought up as a member of the Church of England but was later converted to the Catholic faith. Guy Fawkes went abroad to seek his fortune, becoming a professional soldier. It was when he was fighting with the Spanish army in Flanders in 1604 that he was approached by the architects of the gunpowder plot, who wanted him as their professional adviser because of his knowledge of siegecraft and explosives. The main villains of the piece were Robert Catesby, Thomas Percy and Francis Tresham. Guy Fawkes himself had no part in the planning of the actual plot. Indeed the idea had been put forward as early as 1603 by Robert

Catesby, who had collected round him a band of disgruntled Catholics.

The scheme was originally conceived as a tunnelling operation; a dwelling near the Houses of Parliament was bought and an attempt was made at the excavation which, however, proved unsuccessful. A property which abutted the Houses of Parliament was then purchased so that the gunpowder could be smuggled into Parliament through a connecting passage. The plan went ahead, and in all twenty barrels of gunpowder were smuggled into the cellar under Parliament; large pieces of metal were placed on top of the barrels, so that when the explosion took place it would be even more damaging. Guy Fawkes himself was responsible for this final stage of the plot, the top conspirators having dispersed to the country, leaving Guy Fawkes to face the consequences when eventually he was caught. Fawkes seems a shadowy figure, though he must have possessed a stubborn and defiant nature, for after his apprehension he had to be tortured on the rack before a confession could be obtained from him.

The gunpowder plot caused a great outcry, since it was one of the first of its kind. Nowadays, the Houses of Parliament are carefully protected against anyone who might carry out such a scheme, but in the seventeenth century gunpowder was a relatively recent invention, as far as Europe was concerned, and such dangers had not been fully considered. The furore at the time was understandable, Guy Fawkes seeming to arouse the most extreme horror. The plot was 'an invention so inhuman, barbarous and cruel as the like was never before heard of', 'something which only malignant and devilish Papists could have dreamed up'. Guy Fawkes himself was described as: 'the great devil of all', 'the devil of the vault.'

There is discussion to this day as to whether the whole scheme was merely a romantic intrigue which was bound to fail, or whether it was part of a well-planned *coup d'état*. It is now impossible to be certain, but this does not stop the arguments continuing among the experts. What is without doubt is that Guy Fawkes, who was executed on 4 November 1606, has made himself a name in British history.

The Reichstag fire
The Reichstag fire, like the gunpowder plot, is also a subject for

argument and speculation. Who was responsible? Who actually set the Reichstag on fire? Did it really change the course of German history? The simple fact is that on 27 February 1933 the Reichstag was burnt down. The background however is complex and the story really begins with the defeat and surrender of Germany in 1918.

Following Germany's fall there was compulsory disarmament for that unhappy country, as well as a massive burden of reparations, all of which inevitably left a sense of bitterness mingled with guilt. The Weimar Republic, set up in 1919, struggled on uncertainly; there were problems of galloping inflation which led to bankruptcy and the ruination of the middle class. And in spite of revisions of the treaty the situation in Germany worsened, accentuated as it was by the world economic depression of the early 1930s. It was in such a situation that Adolf Hitler, an Austrian leader of the National Socialist Party, began to manoeuvre himself into power, using terrorism and violence, demagogy, fanatical appeal and relentless propaganda. He was an impressive showman, full of promiscuous promises. The slogans of the new Nazi Party were anarchistic, anti-communist, anti-intellectual and anti-semitic. The party's anti-intellectualism took the form of persecution of the intelligentsia and of the public burning of books. By the mid-1930s it was to become a reign of terror, with concentration camps, firing squads and the secret police. But in 1932 all this was largely to come, for the Reichstag fire occurred during a period in which Hitler was in a desperate struggle for power. The elections had gone badly in 1932 and there was a need to whip up support for the party.

On a bitterly cold February evening a man who happened to be passing the Reichstag in Berlin was suddenly showered with glass. Running to the window he saw inside a dim figure holding a lighted torch, but unable to enter he raced to the nearest police station and the alarm was given. By the time the emergency services arrived the main chamber was in flames. The police forced an entry and in one of the corridors they found a half-naked man whom they arrested. He was a twenty-four-year-old Dutchman named Mirinus van der Lubbe. From that day until his execution by beheading he protested that the fire was his work and his alone.

On that same evening, according to Shirer in *The Rise and Fall of the Third Reich*, there were two separate dinner parties in progress. The vice-chancellor, von Papan, was entertaining President von Hindenberg in the exclusive Herren Klub, while in Goebbels's home Chancellor Hitler was relaxing as the two played music on the gramophone and recounted stories. Later, there was a telephone call from Dr Hanfstaengl; it is recorded in Goebbels's diary that the doctor informed him that the Reichstag was on fire, but Goebbels did not believe him and at this stage did not even mention it to Hitler.

In the Herren club, however, just round the corner from the Reichstag, they were very much aware of what was going on. Papan wrote in his memoirs: 'We noticed a red glow through the windows and heard sounds of shouting in the streets', and then a servant called out that the Reichstag was on fire. Going to the window they saw the Reichstag 'as though it was illuminated by searchlights', with flames and smoke swirling round the dome.

In the minds of all these men, and Goering as well, there was no doubt that this was a communist crime, the beginning of the Bolshevik revolution. Goering was reported by Rudolf Diels, the Gestapo chief, as saying 'every communist deputy must this very night be strung up'.

The following dawn saw the beginning of the arrests. Ernst Torgler, the party's parliamentary chairman, was taken into custody, as were three Bulgarian communists including Georgi Dimitroff. Ernst Thalmann, the party leader was found in hiding a few days later. This frenzy of activity was brought about by Hitler, who had persuaded President Hindenberg to sign a decree, 'for the protection of the people and the state', in which certain sections of the constitution on the protection of the individual and on civil liberties were suspended. It was the first experience for the populace of mass arrests, torturings and beatings. Documents were also published purporting to show that the burning of the Reichstag had been the signal for a communist insurrection. The following year the results of the election declared on 5 March 1933 showed an increase in votes for the Nazi Party to 44 per cent.

Why all the argument and discussion about the Reichstag fire?

It seems quite clear that it was started by a solitary fire-raiser, the Dutchman van der Lubbe, and that the event was promptly used by Hitler's propaganda machine for his own political advantage. Certainly the Dutchman had been a communist, but he had left the party in 1929. The reason for the continuation of the discussion is that vast amounts of propaganda emanated from both the national socialists and the communists. One view is that the Nazis started the conflagration themselves, and indeed Goering boasted in 1942 at a luncheon party that he was the only one who really knew who started the fire because he had done it, a bit of 'evidence' which came out at the Nuremberg trial. But what supporting evidence is there of this?

It was said at the time that there was a passage leading from the president's palace directly to the Reichstag and that it was therefore possible for someone from the palace to have started the blaze. Certainly, as we have already seen, basement corridors are a favourite place for fire-raisers, whatever their particular motives, allowing as they do for both entry and exit as well as providing good ventilation for the flames. In addition basements often contain inflammable material, such as lagging on pipes, rubbish, and so on, another necessary ingredient. Shirer indicates that it was 'through this tunnel [that] Karl Ernst, a former hotel bellhop, took a small detachment of stormtroopers on the night of 27 February to the Reichstag where they scattered gasoline and self-igniting chemicals and then made their way quickly back to the palace, the way they had come.'

If one is to believe this story then one has to accept the almost impossible coincidence that they arrived at the precise moment when the Dutchman was setting to work upstairs with his lighted torch. According to a recent article by Donald Watt this idea of the underground tunnel, and the idea that Hitler's minions set the place alight, is pure invention, a figment of the imagination of Willi Munzenberg. Munzenberg was a communist and a friend of Lenin. He was also a first-class propaganda expert and, in fact, according to Watt, won the propaganda battle by getting Hitler blamed for the plot. His *Brown Book* of the Hitler terror was produced in 1933 and this made much of the secret tunnel between the president's palace and the Reichstag. However, according to Watt it did not exist; there was a maze of locked cellars which would have prevented direct access from one building to the other. In Watt's view, therefore, the tunnel was a fabrication. Another point made by Watt is that Munzenberg's team had to explain van der Lubbe's presence in the Reichstag and it was this that made them invent a sexual liaison between him and Captain Ernst Roehm, who was in fact a notorious homosexual.

In these bursts of propaganda from both sides, only one fact seems to emerge and that was that van der Lubbe had started a fire in the Reichstag. We do not know whether he was paid for his action by one or other party or whether this was a personal act of fire-raising. Was he a 'fire-bug'? Was his behaviour a symptom of severe mental derangement? To determine into which category he is to be placed we would need further information about previous similar activities or some idea of his mental state at that time. This is not forthcoming so we shall never really know what type of a fire-setter he was.

Fire as a weapon of war
The use of incendiary bombs in the Second World War appears to have been as much a weapon directed at the civilian population as against military installations. There is still a good deal of argument about the overall value of these and other techniques of bombing; in purely military terms they can often be expensive in manpower and equipment and may lead to a rise in the morale of the attacked population rather than the reverse. The other side is the moral one; certainly the rights and wrongs of the bombing of Dresden are still debated, but there really seems no authentic debate possible on an even more dreadful weapon – napalm. It was widely used by the United States in Vietnam even though the United Nations had concluded a report, to be published soon, indicating that this weapon produces 'excessive suffering'.

Napalm was first resorted to at the end of the Second World War when the battle was concentrated in the Far East. Apparently General Curtis Le May obliterated large sections of Tokyo with it. Since then technical advances have rendered napalm hotter and tackier so that it sticks more firmly to the victims and makes sure of their death. The type of napalm recently employed in Vietnam was a mixture of benzene, polystyrene and petrol ignited by tiny particles of white phosphorus – 'Willie Peter' in

American air force slang. Dollar for dollar it is the cheapest method ever devised for killing people.

According to Nigel Hawkes:

> Napalm either burns its victims to death or, more subtly, creates such an inferno of flames that all the oxygen in the air is consumed and suffocation follows. [This makes it effective even against people in underground shelters.] The small fragments of white phosphorus often become embedded in the flesh of the victims. There they continue to smoulder, sometimes for weeks, producing agonising pain for which the only remedy is constant immersion in a cold bath.

It is extraordinary that this intensely painful weapon has not caused more public outrage and revulsion. Certainly the television pictures of a particular child, a napalm victim in South Vietnam, were among the most gruelling and grisly of the whole terrible war.

Aggressive fire

Aggression is probably one of our deepest instincts. Everybody has aggressive impulses but it is hoped that in the course of their upbringing they will learn to control and channel them. Anthony Storr writes in *Human Aggression*: 'It is not only a valuable part of the human nature, but also an essential ingredient in the structure of society; and it is only when the aggressive drive becomes blocked or frustrated that it becomes objectionable or dangerous.'

We tend also to think of animals as highly aggressive creatures; thus we talk of 'nature red in tooth and claw'. This is a misinterpretation of the Darwinian view of the survival of the fittest because the animals do not kill one another out of aggression but are engaged in a struggle for food. When there is overcrowding, however, the situation is quite different. Recent studies by ethologists on both monkeys and rats have shown that overcrowding may bring about aggressive behaviour, and in man sociological surveys indicate that the same is true; in areas of high population and poor living conditions aggressive feelings are uppermost. Violent acts may then occur because aggression, no longer controlled in any rational way, is turned into hatred and hostility.

We tend to think of the Dark and Middle Ages as times of marauding armies who looted, raped and burnt, as for instance the early Scandinavians who attacked villages and hamlets on the east coast of England and then finished them off with fire. But this behaviour is still very much with us in other parts of the world, as in West Pakistan during the civil strife in 1971, where villages were razed, their occupants killed, and their bodies cut into pieces and incinerated.

The hot overcrowded suburbs of American cities are also places where the desperation of oppressed people is seen and where aggression wells up and violence breaks out. Then fire is a weapon. As Jeremy Campbell of the *Evening Standard* put it in his report from Brownsville, a suburb of New York, 'It is high noon, and black smoke is billowing out of a tenement house on Satter Street in Brownsville, a residential district of New York. From a broken window on the top floor, mingled with the banshee scream of the police siren can be heard the soprano panic of a woman. The morning is all soot and sweat and blasphemy.' The Bristol Street fire brigade answers an average of fifty calls every day of the year and in these desperate suburbs it is not always greeted with relief but often by hails of bricks, bottles and sometimes bullets.

Fire is used by desperate people when they have no other weapon, and always by anarchists as one of their prime agents of destruction. The 'angry brigade' in England, for instance, has been responsible for blasts and blazes all over the country; in a similar fashion, the 'weather men' in the United States, a turbulent, anarchistic group, have set off bombs all over the United States. They have viciously attacked the establishment, sparking off political controversy, rioting, bombing, looting and arson.

So often, it is the general feeling that by ordinary political action citizens are unable to bring about any change which causes aggression to turn to hatred. Northern Ireland is another place where arson, bombs and bullets reign. In this state of violence some surprising psychiatric statistics have been reported. The suicide rate has fallen, and the number of people with depressive illnesses in general has also decreased markedly. Is there any relationship between these statistics and the aggressive behaviour seen daily in the streets?

Aggression and depression
To some degree everyone has experienced depression. The death

of a loved one, failure in business, and disappointment in love can all lead to sadness, if only briefly. It is true also that even the mildest form of depression changes behaviour – perhaps the person does not want to see his friends, or to go out; or when confronted with situations in which he must act he becomes irritable, loses his temper, and shuns the very people who might be able to help him. But this type of withdrawal from social contact can become so severe that the disturbance amounts to a depressive illness requiring specialist treatment. In most cases, because of the individual's experience of comfort and succour given by parents, especially by his mother, during childhood, the individual has a built-in mechanism which helps him over periods of loss and disappointment. He is able to counter the mood of sadness and depression and slowly return to a usual state of well-being. As Anthony Storr puts it: 'He could be said . . . to carry inside himself an inner source of love which is independent of external vicissitudes. There seems every reason to suppose that self-confidence in later life is based upon the infant's earliest experience of his mother.' The psycho-analyst Melanie Klein explained the infant's inner sense of security as 'introjection of a good breast'. She also saw the child as biting this breast and scratching it, showing how in her view the love-hate interaction is built up even in a young child.

The depressed person is in a state of inaction, still having inside himself aggressive energy, learnt as a child and nurtured through life, which has to be disposed of somehow. If he is unable to do this the powerful forces build up and can lead to suicide as, unable to escape, they become directed inwards. We see this in individual terms in the depressed patient. However, we also see it collectively; when there are no means by which a group can channel its aggressive energies violence flares.

But all is not as clearcut as this, since unchannelled aggression arising in the course of a depressive illness can turn outwards as well as inwards. In a recent study, 'Murder Followed By Suicide', D. J. West shows that one-third of the murders committed in England are followed by suicide. 'The intimate connection between self-destructive and aggressive tendencies emerges clearly from many incidents in which offenders' intentions waver uncertainly between murder and suicide.'

We have already mentioned that the suicide rate in Northern

Ireland has fallen. In a recent study, H. A. Lyons has shown that during the spells of violence in the years 1969-70 there was a decline in depressive illnesses as well as a decrease in the number of suicides in Belfast, though not in the quiet rural areas. The fall in the suicide rate was as much as 50 per cent but this was coupled with a noticeable rise in the murder rate as well as a big increase in rioting, arson and bomb explosions. Hence it appears that if a person has access to some means of venting aggressive feelings he is less likely to suffer from depressive illness.

An aggressive act such as arson does not arise just on the basis of an individual's inability to deal with his violent feelings. It arises in the context of poor social conditions in general. This, as we have mentioned already, has been examined in relation to animals and it is found that, in particular, overcrowding is one of the trigger factors. The areas in Belfast in which the troubles have arisen are largely those with bad housing, poor facilities for recreation; and, indeed, the whole of the province of Ulster is afflicted by high levels of unemployment, not to mention the problem of conflicting religious beliefs. Another factor noticeable is the particular problem in junctional regions where Catholics and Protestants live side by side. Indeed one of the most frightening aspects of the conflict was the burning of a row of Protestant houses next to a Catholic area. It was not an act of arson. The people left of their own accord, taking their belongings with them, deliberately burning down the property so it could not be occupied by the opposing group. There is further evidence that we are witnessing an effect of the territorial aspects of aggression, something experienced and written about by the ethologists: the erection of barricades. The areas round these become then the scene of violent activity, shooting, bombing and arson.

Street violence as Professor S. L. Washburn points out is almost the prerogative of the young male.

> Throughout most of human history society has depended on young adult males, to hunt, to fight, and to maintain the social order with violence.

However, in Belfast the situation has been somewhat different in that older men have taken part and women have also been involved. In fact during earlier rioting when C.S. gas was used it was noted that 35 per cent of the gas casualties were female.

The women also are seen in the streets encouraging the men, and, indeed, their presence may make the males strive to greater violence since this adds an element of sexual rivalry and demands a higher degree of aggressive maleness than one might otherwise expect.

There are two other aspects of the type of conflict in progress in Northern Ireland. The first is that an aggressive minority continue and increase criminal activities at the same time and quite independently of the street violence and political struggles. For example in Northern Ireland there has been a marked rise in armed robbery. This has presented such a problem that the IRA themselves have had to set up their own courts and mete out punishments to offenders. The other aspect is the way violent movements attract to themselves aggressive and sadistic members of the community. They are people who, up to the time of the troubles, had kept their perverse instincts under control. Now they have a relatively acceptable outlet. Men with sadistic tendencies have obviously been involved in Northern Ireland as they were in other para-military movements elsewhere. Anthony Storr points out that this sadistic aspect of man is something that marks him out from the animals. They fight and kill each other but rarely, as far as we can judge, take any particular pleasure from this activity.

> Burn, burn, burn the soldiers
> Burn, burn, burn the soldiers
> Burn, burn, burn the bastards
> Early in the morning.
>
> Children's song from Belfast 1971

Just how important the learning of overt aggressive behaviour is can be seen in Northern Ireland. Television films show children regularly in action against the troops and the children's song from Belfast indicates quite clearly their attitude to the soldiers. Unfortunately the proliferation of violent attitudes is not limited to relatively small parts of the world, because recent studies have shown that very young children shown overt aggressive acts on television quickly imitate them with their toys and their peers. We know something of this at a personal level from Tony Parker's *The Courage of his own Convictions*. 'Violence is in a way like bad language – something that a person like me has been brought up with, something I got used to very early on, as part of the daily scene of childhood, you might say. I don't at all recoil from the idea, I don't have a sort of inborn dislike of the thing like you do. As long as I can remember I have seen violence in use all around me; my mother hitting the children; my brothers and sister all whacking our mother, or other children; the man downstairs bashing his wife and so on.' Ireland has been the scene of aggressive outbursts almost for centuries. It seems that the whole area has been nurtured in a climate of aggression and hatred. After all these years it is difficult to see how this tide can be turned. Violence in general like fire can be a self-perpetuating process.

Chapter Seven
Ordeal by Fire

On 2 January 1969 a young Czech student ignited the petrol which he had poured over his body. Flames soon enveloped him. He was taken to hospital desperately ill, to survive only a few days. This deed was done in Wenceslas Square, Prague, soon after the Russian invasion of Czechoslovakia, and by it Jan Palach has made a mark for himself in history. It was a brave act, and not only because it is reminiscent of a similar sacrifice of another Czech who felt that this could be the only response to a desperate situation.

Huss and Palach
Some five and a half centuries ago, John Huss, a teacher at Prague University and a follower of the Englishman John Wyclif, was burnt at the stake. Wyclif, it will be remembered, translated the Bible into English and wanted everyone to read it in his own language. So also did John Huss. And for this belief he chose to die, for if he had recanted he would have been saved. Instead he left a prayer:

> Seek the truth
> Listen to the truth
> Teach the truth
> Love the truth
> Abide by the truth
> And defend the truth
> Until death.

The death of John Huss in 1415 spurred the Czechs into action. Led by Jan Žižka, who was a gifted tactician, they managed to beat back the Catholic armies, and for over two hundred years Protestants in Czechoslovakia succeeded in maintaining their independence from the domination of Rome.

Jan Palach, a philosophy student, was known to his colleagues and professors in Prague's Charles University as a rational thinker. He took deliberate action, after the Russian invasion, to rouse Czechslovakia from its passivity. He was at first successful, for following his action hundreds of thousands of people moved in procession through the centre of Prague to honour his memory. Others felt that a more positive step should be taken. Some youngsters attempted to set in motion a wave of violent protest. Some even wanted to follow in Jan's footsteps by setting themselves on fire. However the authorities were alerted and the uprising was quickly quelled. In the long term the action of Jan seems to have left little impression on Czechoslovakia itself, but it captured the imagination of the world press and is certainly still remembered. He was buried in Prague's Olsany Cemetery and the grave, though at first it became a place of pilgrimage, was in February 1971 desecrated by the authorities who took the bronze relief from its stone setting and melted it down. Obviously they feel the memory of Jan still lingers and attempts must be taken to erase it.

So far in this book we have discussed people who have used fire to further their own ends. Their action, though outside the generally accepted norms of behaviour, is understandable, for their motives have a rational basis even though the action itself is not, and cannot be, condoned by society in general. They kindle fires for profit or to cover up crime. We have also seen fire used in the violent release of tensions in situations where ordinary political means of expression did not exist. The people involved in this 'abnormal' use of fire are not apparently mentally deranged except in this one respect. However, in this chapter we are examining a new aspect of the abnormal use of fire, suicide by burning, which we must try to understand.

Some might say that no sane person could perform this act. So what evidence is there about the mental state of Jan Palach at the time of his self-immolation? Was he or was he not seriously mentally deranged? His girl-friend Eva Bednarihor saw him in the hospital over the three last days of his life, and Jan told her that he did not want others to follow his example; thus he obviously knew what he had done and how it would finish. His mother, too, said of him that 'he always spoke the truth and could not bear injustice, always a good Czech'. We also have a handwritten letter by Jan himself about his act of self-immolation, the text of which was published in full in *The Times* on 11 February 1969:

> In view of the fact that our peoples find themselves in a difficult situation and have reached the brink of resignation with regard to their fate, we have decided to register our protest and to arouse the conscience of the nation in the following manner: our group is made up of volunteers who are prepared to burn themselves in the literal meaning of the word for freedom and democracy in Czechoslovakia. I have the honour to draw the first lot, and thus the right to pen this letter and to become the first torch. Our demands are: the immediate abolition of the censorship and the prohibition of the circulation of *Zpravy*; the resignation of some politicians who do not enjoy the confidence of the people.

He then went on to say that if the demands were not met there would be a general strike, and new torches would flare; and he signed himself 'The Torch'.

On the basis therefore of this evidence Jan does not appear to have been actually mentally deranged; but perhaps we can understand his motives if we look upon them as similar to other political fire-raisers. The natural aggression they have within themselves is transformed by the circumstances in which they find themselves into hatred and violence because it is unable to find any other release. Jan felt unable to do any more through ordinary political action. There were no physical means to fight against the steel of the Russian tanks. Therefore he chose a spectacular death to draw attention to the censorship and oppression in his country.

We may picture Jan, then, as a serious, well-intentioned young man, expressing his feelings of desperation through self-immolation.

This is, however, not the opinion of A. Alvarez. In a study of suicide, *The Savage God*, he points out that there is a supernatural quality in the action: 'He is performing a magical act which will initiate a complex but equally magic ritual ending in the death of his enemy.' It is as though in destroying himself Jan destroyed his persecutors, a type of reasoning which Alvarez compares to the magic of primitive societies.

Vietnam and suicidal protest

> I cried a long time tonight
> Because a Mr Beaumont
> sent me a letter
> praising me
> and thanking me for the things I do
> in my struggle for peace. . . .
> But do you recognise his name?
> His wife
> Florence
> burned herself to death
> in front of the
> New Federal Building
> in Los Angeles.
> She poured
> gasoline
> on her clothes
> and lit a match to herself
> to illuminate the dull fact that
> children
> children
> little children
> are being burned to death
> in yellow fire. . . .
> She gave a gift of
> her whole self
> to remind the rest of us that
> no matter what we're doing,
> it's not enough.

Thus wrote Joan Baez in *Daybreak*. Suicide by burning began in Vietnam in 1966 and this method of protest has flared spasmodically ever since. Of interest in this connection is the part played by fire in Buddhist writings. There is for example the 'Fire Sermon' (translated by Henry Clark Warren):

And with what are these on fire?
With the fire of passion, say I, with the fire of hatred, with the

fire of infatuation; with birth, old age, death, sorrow, lamentation, misery, grief, and despair are they on fire.
The ear is on fire; sounds are on fire . . . the nose is on fire; odours are on fire . . . the tongue is on fire; tastes are on fire . . . the body is on fire; things tangible are on fire . . . the mind is on fire; ideas are on fire . . . mind-consciousness is on fire; impressions received by the mind are on fire; and whatever sensation, pleasant, unpleasant, or indifferent, originates in dependence on impressions received by the mind, that also is on fire.

Fire, then, is an all-pervading force in the mind of Buddhists, but that is not the whole explanation. It must be considered in relation to two main aspects of the Buddhist philosophy, first that after purification transmigration takes place, namely that the person is reincarnated. The other aspect is put as follows by Howard Chambers, general secretary of the Buddhist Society: 'The only relation it [suicidal protest in Vietnam] has to Buddhism is that it would be thought better to kill oneself than somebody else. Otherwise there is no tradition of this particular form of suicide or of suicide generally forming part of the Buddhist religion.' Be that as it may, it was protest-suicide by Buddhist monks in South Vietnam, setting fire to their petrol-soaked clothing, that became adopted as a technique the world over. Joan Baez describes above one such suicidal protest in the United States. Jan Palach in Czechoslovakia followed in the Buddhists' footsteps, and after his death there was an outbreak of suicide by fire in northern France. During 1969 and 1970 at least six adolescents attempted this, often choosing the anniversary of Jan's death. There has also been an incident in Italy. One teenager left a note which read:

Against violence, against war and the destructive madness of man I have decided to die.

It is recorded in *The Times* of 20 March 1971 as follows:

A young man poured petrol over his clothing and set himself on fire by jumping on to the eternal flame at a Genoa memorial today. This was a protest, according to a letter he left, against the war in Vietnam.

It seems perhaps surprising that such a horrible method of defiance should be contemplated; is it conceivable that, as in so many other things, there is a fashion in suicide? Surprising though it may seem, this is the view taken by Professor E. Stengel, who has studied suicide in depth. He believes it is not the suicide itself but the technique used which is influenced by publicity. Alvarez, too, looks at this point of view:

> The periodic epidemics of suicide are not dissimilar to the first athlete who broke the barrier of the four-minute mile. There are examples of fourteen wounded soldiers who hanged themselves on the same hook in 1772. This was at the Les Invalides Hospital in Paris. The epidemic stopped when the hook was removed. Thousands of Russian peasants burnt themselves to death in the seventeenth century. They believe that 'anti-Christ' was coming. In Japan in 1933 hundreds threw themselves into the crater of the volcano Mihara-Yama; access to the mountain was closed and the suicides stopped. In Chicago there was a suicides' bridge and to stop the epidemic it was demolished. It seems therefore that a particular technique, if adopted by someone and then publicised, the phenomena may spark off a conflagration.

Fashion is only one aspect of suicide; the cultural background is of even greater importance. In India, for instance, *suttee* was a common occurrence. With the husband burning on his funeral pyre, the wife throws herself in desolation on the flaming embers. This can be regarded as no more than a despairing reaction to a husband's death, a reaction which is extreme but not unnatural during the period of mourning. But there is more to it than this. The act has to be considered in relation to the cultural setting; the widow may realise her new and lowly status as a widow and this, together with her natural grief, leads her to suicide. *Suttee* is largely a thing of the past, but similar acts still occur in India from time to time. In 1969 a young man in New Delhi gave his wallet to his friend and then threw himself on to the sandalwood pile in which his guru was being cremated.

Saints and martyrs
Fashion and culture, then, make their impact on suicide patterns. Another aspect which may now usefully be looked at, if only to indicate the complexity of human motivation, is the changing pattern of suicide technique throughout history. In Roman times, for instance, Christianity was outlawed by the authorities because the believers refused to make the obligatory sacrifices to the gods.

They were punished by being fed to the lions, a fate which, far from terrifying them, they welcomed as being the means to glory and salvation. To quote Alvarez again: 'The persecution of the early Christians was less religious and political than a perversion of their own seeking. For the sophisticated Roman magistrates Christian obstinacy was mostly an embarrassment: as when the Christians refused to make token gestures towards established religion which would save their lives.'

We see here another aspect of suicide, which applies also to those whose choose suicide by fire: submission and pleasure in submission, a state to which the term masochism is applied by psychoanalysts. Karl Menninger in *Man Against Himself* sees ascetic actions and martyrdom as a self-destructive pursuit. He restates the belief of Freud in the 'death instinct' and puts forward the view that this may be, not entirely a passive, but an active process, and further suggests that these facets of personality are present in every one of us. He also cites examples from earlier times when thousands of Christians are said to have leapt in frenzied joy over the overhanging cliffs so that the rocks below were reddened with their blood. It seemed that the believers were so aware of the reality of heaven – their eternal home – that life itself was unimportant. John Donne observed:

> [It seemed] that those times were affected with a disease of this natural desire of such a death. . . . For that age was grown so hungry and ravenous of it [martyrdom], that many were baptised only because they would be burned, and children taught to vex and provoke executioners that they might be thrown into the fire.

This element of provocation of the authorities still persisted into the Middle Ages. Joan of Arc could easily have signed a document recanting her heresies but she would not, and she further defied her captors by dressing up as a man and attempting to go to mass in such a garb. These and other actions led to her death at the stake. One does not want to belittle her character, any more than one wants to minimise the sacrifice of Jan Palach; one merely wants to underline the fact that acts of this kind are in the last analysis acts of extreme passivity because of the dominance of masochistic elements in these indidivuals' personalities.

There are other examples. Take for instance the case of Bishop

Hugh Latimer, an English bishop who lived from 1485 to 1555, not apparently a great theologian but one who was always involved in controversy and religious dissent. He was an important figure during the Reformation but when Mary Tudor came to the throne he was at once in danger. Now the element of provocation; Latimer stood out against this backlash of the Reformation. There was trouble, and finally he was burnt at the stake with Bishops Cranmer and Ridley. Tradition has it that even when the flames had barely reached Latimer's feet he stretched down his hands into them – the masochistic element? Bishop Latimer's own words just before his death have been preserved: 'Be of good comfort, Master Ridley, and play the man; we shall this day light such a candle by God's grace in England as I trust shall never be put out.'

The motivation of Latimer, as in all such acts, is complex, but masochism must play a part as must magical thinking, for by his death he felt he would in some way convert his enemies.

Suicide by fire
There are other deaths from burning which seem to contain none of the political or religious overtones of the examples we have so far examined. These are those individuals who choose fire, heat and burning as a method of killing themselves. Menninger quotes examples of people throwing themselves into boiling vats of liquid, swallowing pokers, and clutching hot stoves. I myself was able to collect without difficulty at least seven newspaper clippings relating to suicide by fire in 1971. Three of them were elderly people, and only one was a young man, said to be a 'hippy'. These fortunately relatively uncommon instances are known to the forensic pathologist, but generally there is little information available about their previous lives, other than that they were lonely and desperate people. One case quoted by Professor Polson, the forensic pathologist, concerned a young man of twenty-one. His life was totally disorganised, he had been employed in many jobs, and he had previously been in trouble because of acts of gross indecency. He had attempted suicide with drugs and had received psychiatric treatment on several occasions. Finally he bought some petrol and burnt himself to death. The psychiatrists who looked after him before regarded him as suffering from a severe schizophrenic illness, and it may

be that in his turmoil hallucinatory voices had haunted him and led him to this bizarre death. Others, too, in severe depressive phases have resorted to these terrible means.

Clearly suicide by fire has many aspects. We still do not understand clearly, and perhaps never will, why particular individuals turn to this method.

PART THREE

THE MOTIVELESS FIRE-SETTERS

Chapter Eight
The Incendiarists

So far we have examined several types of fire-setter who in their own ways have had what could be called valid motives. In some cases individuals have used fire for their own profit, or to cover up crimes; in other cases groups of people have employed fire as a political weapon. We have also looked at people who have killed themselves with fire, and at least some of these have had political motivation. In this chapter, and for the rest of the book, quite a different aspect of fire-setting is discussed. These are the fire-setters who cannot claim the kinds of motive already mentioned, who apparently start a blaze as an end in itself. This is a difficult field, but one which is of the greatest importance, because if we can trace the underlying mechanisms which lead to this particular behaviour, and which put people and property at risk, then we can perhaps learn how it could be controlled and how these individuals can be helped by a psychiatrist.

To the ordinary person the incendiarist is quite bizarre, utterly misguided and completely mentally deranged, and certainly this is one way of regarding his behaviour. Closer examination, however, reveals motiveless fire-setting to be more complex, as by no means all fire-setters seem to be mentally disturbed. It is often useful to break down a particular behaviour pattern into various groups in order to understand the cause of it. Take the matter of murder; all murderers do not give the same reasons for their acts; and as we shall see this applies also to fire-setters. To us they appear motiveless, but the fire-raisers themselves often give reasons for their acts. Nevertheless, it remains true that these reasons are often inappropriate, or appear quite out of proportion to the deed committed. (For example, a man is dismissed from his job as a car mechanic and in a fit of pique he kindles a blaze at the garage.) And in any case it is difficult to evaluate the reasons given by fire-raisers, as sometimes they are simply lying, while at other times the explanation is concocted after the act. As we shall find out in the case of the 'fire-bug', he is often in a state of confusion and quite unable to give a rational explanation after his series of blazes.

In the earlier chapters we examined how our 'motivated' fire-raisers might set one or many fires. Van der Lubbe apparently started only one blaze, whereas Leopold Harris and his gang were responsible for many and the members carried out their activities deliberately and repeatedly. Similarly, in the group without clear motives, there are some individuals who set only one fire in a lifetime, but much more common and dangerous are the 'repeaters'. It may be too that those who set a single fire are stopped by the intervention of the law, otherwise they would more than likely have continued their activities. Interesting in this connection are the well-documented instances of people who set a fire or two when they are teenagers and then not again until they are much older. At that time, presumably as a result of senile changes in the brain, their ability to control their behaviour is lessened and they revert to fire-raising.

Are there characteristics of age, sex or intellect which mark off the motiveless fire-setters from the rest of humanity? Not surprisingly there is a preponderance of one sex – the male. Fire-setting, like most other forms of anti-social activity, is much commoner among men than women. Further, there is a peak incidence in males in the late teens and early twenties, whereas in women an even spread over all age groups is found. As to the matter of intellect, there is an abundance of those at the lower end of the scale of intelligence though still within the 'normal' range, while a few highly intelligent people also turn to this form of activity (recently a French professor of psychology started a dangerous fire in the Riviera). But there is also a well-defined group of mentally subnormal who start fires. In answer to the question, are these people mentally deranged, the simple but incomplete answer for the majority of cases is No. Apart from their abnormal attitude to the use of fire, there is no characteristic mental disorder. There are nevertheless some fire-setters, as we shall find out, who do suffer from schizophrenia or from some other form of severe mental disorder. Again, most fire-setters are not sexually perverted, but there are those who are, as we shall recount later. Vagrants form another group who show particular patterns of fire-setting, and then there are the rare instances in which firemen themselves turn to incendiarism.

It is never possible in a discussion of human behaviour, especially of the 'abnormal' kinds, to give clearcut reasons or to form watertight classifications for the various types of action, and in our discussion of the 'motiveless' and the 'motivated' fire-raiser we encounter just this difficulty. There is always some overlap, and certainly when consideration is given to fire-setting, such 'grey areas' are found in particular abundance. One example from this 'grey area' of the fire-setter is the one who says that he has been offended or slighted in some way or other by a neighbour and so sets light to the neighbour's house. This is an example of the so-called 'revenge' or 'hate' fire.

A blaze may thus be set in a fit of passion, after a domestic fracas, following dismissal from work, or as a result of some real or imagined slight. Such incidents generally follow soon after the event, a few hours or days later, though fires relating to some deep-seated feud may be set even years after the crucial incident which was the stimulus. Usually blazes of this type are set only once in a lifetime, but of course the fire-bug who sets many fires for no particular reason may at some period in his life, as a result of some feud, set a fire which shows some semblance of a motive. Women seem particularly prone to revenge fires, as in the case of the woman who, when the man she had been living with left her, paid her teenage daughters 25 p to set his new lodgings on fire. This particular incident is unusual in that she did not undertake the act herself.

Sometimes hate is directed against a business rival. An unusual case of this kind occurred in Britain recently when a man firebombed a pop radio station located on a ship in the North Sea. This radio station was a relative newcomer and was competing with the other pirate ships which had been on the air for a considerable number of years. As a result of this incident a number of arrests were made and it became clear that the fire-bomb incident was intimately connected with the business rivalry between the stations.

In the same year a similar situation arose in the East End of London. This time the rivals were in the cab hire business; one set the other on fire but this time the damage was much more serious than had occurred in the pop ship incident in which relatively little damage to property resulted. Two died in the East End blaze.

Hospitals, like churches and schools, are frequent targets for fire-setters. A recent example was of a fire at the Evelina Hospital for Sick Children in South London, set by a member of the domestic staff who felt that she had been wrongfully dismissed. Fortunately, little damage was caused. The same was not true in another hospital fire in 1971, this time at the Radcliffe Infirmary in Oxford. It was estimated that a quarter of a million pounds worth of damage was done, mainly because medical journals, textbooks and irreplaceable documents were destroyed. One might be forgiven for wondering whether a medical student who had failed an examination was responsible, or perhaps a disgruntled patient. No one however was caught, and since then a new lecture theatre has also been gutted. This was also the case at my own hospital, the London Hospital, where a series of fires at weekly intervals occurred until a disastrous blaze enveloped the X-ray department, causing such damage that it was closed for almost a year. The culprit was never found.

Another group of fire-setters could be labelled as 'the heroes'. They are moved primarily by vanity, and can be summed up in the words of La Rochefoucauld: 'There are heroes in evil as well as good.' This group tends to be older than other fire-setters. They are of poor physique and of dull intelligence, as well as being impulsive and immoral, often responsible for assaults of various kinds including rape. Their immature nature is revealed through their alcoholism and promiscuity and their hankering after heroic games. These people set fires and then call the fire brigade, remaining behind to help people to be rescued from burning buildings, even assisting the firemen to roll out the hoses. Like most of the fire-setters who have no particular motive, these people have never achieved anything during their lives; they feel that they have not fulfilled their expectations and so they change the 'course of history' to obtain success. Sometimes nurses are responsible for such fires, and as the blaze spreads they are seen carrying the babies from the flames. Nightwatchmen too have been incriminated for this kind of fire. They are always first on the scene and can raise the alarm as well as 'rescuing' the potential victims. Though uncommon, one such case was reported in 1972 when a man rescued four children from a blazing house in South London. His bravery was commended and he was even awarded a certificate. Subsequently it turned out that the fire had not been accidental, and later he was convicted for stealing from both a hospital and a local store.

There are those too who set fires for thrills. They enjoy the blaze and are delighted by the excitement which the arrival of the fire brigade brings. Sometimes teenagers at a party become wildly excited by drugs, alcohol and the music. In the course of their pursuits the house is set on fire, either by accident or deliberately. In the case of just such a party, reported in the *Evening Standard* of 3 June 1972, a house in Bromley, Kent was virtually destroyed because during the blaze the gas main exploded and brought the roof down. Another recent conflagration, probably in the same category, resulted in the destruction of a building and stock estimated at £34,000. Apparently it had been started by two hippies who dabbled in black magic and the occult. They were paid £500 for starting the blaze, and it transpired that the two youths had been taking LSD repeatedly over a period of months and in this state of complete mental confusion they were persuaded to carry out the act of arson. Perhaps they wanted the excitement, or perhaps they needed money to buy more drugs.

These fires set by thrill-seekers are not particularly common but clearly they do occur. They represent one group whose motives are obviously inadequate compared to the havoc they create. In subsequent chapters we will find other groups who set fires for various reasons which they consider important, but which to us seem quite pointless. However, before we continue, certain points of terminology must be clarified.

Generally I have used descriptive non-technical words which seem appropriate since they do not prejudge the issue; for example I have used the word 'fire-raiser' or 'fire-setter' rather than 'arsonist'. Another term used is 'pyromaniac', which literally means fire mad, a word which conjures up many images but nevertheless seems to me to have a slightly old-fashioned ring. In fact pyromania has been used to describe people who merely enjoy fires and seeing the firemen in action, though others have used it to describe practically any person who carries out repeated acts of fire-setting; because of the confusion that may arise I have rarely employed it. The other term which springs easily to the pen is 'fire-bug'. However, I have used this not just for any fire-setter but for a fairly specific category, who will be discussed in a later chapter. Briefly, the 'fire-bug' has an irresistible desire to set fires. He can give no explanation for his behaviour, and even a careful psychiatric examination can usually find no motive. His fire-setting is unprepared, its main characteristic being its repeated nature. He may start many blazes in one evening and then none for a long period. Because of the great number of fires he starts it is inevitable that some get well and truly out of control and become really dangerous, not only because they destroy property but because they may cause loss of life.

Fire-setting when no convincing motive can be given by the perpetrator is a complex condition. In the next chapter we shall examine children who set fires, because through their actions and behaviour we can see how the seeds of the disorder which leads to fire-setting in adult life are sown.

Chapter Nine
Children as Fire-Setters

Children quickly gain experience of fire, whether at the fireside or through attending a bonfire, or in symbolic terms in stories of dragons breathing flames. It is said that the first colour a baby recognises is red, and certainly an infant will sit and gaze with fascination into the flickering flames in the grate, though one cannot tell whether this is a purely physical attraction or whether it constitutes a stimulus to fantasy. A German neuropsychiatrist, Wagner-Jauregg, felt that this attraction which fire holds forms the basis of the impulse to arson in children; and indeed it does appear that fantasies of fire-setting are extremely common in children and are not confined to those who later turn to fire-setting. Certainly these fantasies are commoner in boys and perhaps this ties up with the almost universal desire boys have to be firemen.

Dreams and fantasies of fire and fire-raising occur in many disturbed children and adolescents, but before we examine those who actually turn to these pursuits we must look more generally at the types of psychiatric disturbance which characterise these age groups.

Psychiatrists usually divide children with mental disorders into two main groups. The first is labelled 'neurotic'; these are the children who are anxious, have nightmares, wet the bed, suck their thumbs and bite their nails. As they get a little older they develop 'school phobias'; that is, they refuse to attend classes because it makes them too anxious. Other children are thought of as suffering from 'conduct disorders'; on the whole these children are devoid of anxiety, and the disturbances they show can loosely be regarded as 'anti-social'. They include lying, stealing and truancy from school. It is this latter group which tends to produce the fire-setters.

How many children set fires? In order to determine the magnitude of the problem an American child psychiatrist, Lauretta Bender, made a study in conjunction with the chief fire investigator of the New York boroughs of Manhattan, Bronx and Staten Island in the period before the Second World War. It was found that charges of incendiarism were preferred against 269 persons altogether, the majority being men. Only 72 of these 269 were under the age of sixteen; the group included one girl and ten boys under the age of ten years. On the whole, therefore, very young children seem to set fires comparatively infrequently. Nevertheless, in England there has been in recent years a startling increase in fire-setting by this age group. In 1958 there were only 58 cases but this number had swelled to 156 in 1968. More recently still, in the borough of Brent in London serious fires were caused at four schools in 1972, the damage being estimated at £50,000. A youth club was also destroyed and a new extension to a secondary school flooded. The cause of this increase is not immediately apparent; it may be related to drug-taking, which has increased during the last decade, or to a generally more violent culture with university strikes and demos, a type of activity which has percolated down to school attenders. There have been school strikes and, on the whole, a climate of dislike of authority in any form has grown. This increase in school fires is also in line with the overall rise in the rates of 'malicious incendiarism'.

Lauretta Bender and her colleagues also carried out a detailed study of sixty children who had a history of fire-raising and who were seen in the children's wards in the psychiatric division of New York's Bellevue Hospital during the years 1937 and 1938. The total number of children and adolescents admitted during that time was 1,755, which again confirms the rather limited numerical importance of this disorder for child psychiatrists. The ages of the fire-setting children formed two separate groups, thirty-five between the ages of six and eight and twenty between the ages of eleven and fifteen. Generally they were not particularly outstanding intellectually and five were classified as mentally subnormal. One striking feature of the six-to-eight age group was the occurrence of what has been called the 'minimal brain dysfunction syndromes'. Such children, though they do not suffer from major neurological disorders such as cerebral palsy, have a variety of minor brain disturbances which lead to clumsiness, poor hand-eye coordination, and learning disabilities particularly in reading.

The older ones, between eleven and fifteen, were often referred to the psychiatrists because they had been caught starting fires. These children were characterised by a rather nonchalant attitude to their fire-setting activities; they neither expressed any anxiety about their behaviour nor did they seem to have any regrets. Indeed they appeared to regard it in an unemotional, matter-of-fact, non-involved way. As Macht and Mach noted, fire-setting was not regarded as 'alien' to their way of life. It was planned and carried out away from home, and there was an obvious enjoyment to be gained from the noise and excitement as the fire engines arrived and the men got to work. Even the damage done, which in some instances ran into thousands of dollars, did not cause concern. There was a tendency for some boys to operate in pairs, often a more active aggressive youth associating with a more passive one. This suggested to Lauretta Bender that there may possibly be some underlying homosexual aspects to the association, although there did not appear to be any overt sexual activity.

The younger group, on the other hand, were often referred to the hospital for behaviour disorders other than those related to fire-setting. They tended to steal, be over-active and destructive, run away from home, and play truant from school. These children were quite different from the adolescent group in that they did not collect materials suitable for incendiary acts – matches, candles and the like. The young ones enjoyed building fires in the streets, alleyways and open spaces, the sorts of places in which children usually play.

I myself have a half-remembered fire incident when I was about two and a half years old. It was a Sunday teatime in a country farmhouse. We had a neighbouring farmer and his family visiting us, and the mothers were making tea while the fathers inspected the crops and stock; my friend, a little over five, and I were left temporarily to our own devices. I remember vividly a blaze in the hearth, before which I was alternately warming my hands and clapping them, shouting with glee at the spectacle. This drew the attention of the mothers and our bonfire was extinguished. As far as I remember the five-year-old had collected some newspapers and set fire to them. No damage was done but I dare say he received a few wallops.

The young fire-raisers studied appear to have quite a different reaction to mine at the age of two. They are terrified when the fires that they have started get out of control, and help is quickly sought. They are overwhelmed by the fire, nor do they delight in the excitement which is part of the game for adolescent fire-raisers and indeed adult fire-raisers also.

Lauretta Bender observed that the scene of the blaze for many of the younger ones was in and around their own homes. The fires were not usually serious and the children became very anxious when questioned about them later. Denial of any interest or knowledge of a particular incident involving fire was common, even though information implying the contrary was available to the questioner. These children often had dreams of witches, bogeymen, and ghosts, as well as a great deal of fantasy about destroying parents and older brothers or sisters.

Incendiarism in schools or public buildings was the prerogative of the adolescent fire-raiser. However, the more mentally deranged or mentally subnormal the child was, the more likely was it that the family could not cope with his other problems apart from that of fire-raising, and then placement in a boarding school or other institution often followed. Such children then became even more disturbed and perplexed, because attachments to staff tended to be more unstable than those even in a bad home. It was these children who often later became involved in incidents of fire-setting in their own particular school or institution.

One fire-setting 'game' played by two adolescents reached the headlines in Britain in 1971. A series of fires was started in the homes of Pakistani immigrants in Bradford, and as all the attacks had been made on the homes of one racial group the police were concerned as to whether they represented some form of racial or political violence. However, it became clear that this was not the case and that it was almost certainly two Pakistani boys who were responsible for the incidents. The attacks started on 15 June 1971 and altogether there were thirteen fires in the immigrant area of

Bradford known as Manningham. The blazes were started by dropping a firelighter or pouring paraffin through the letter box. It was on 16 October 1971, as a result of one of these incidents, that three children died. Whether this was the total deaths caused by these boys will never be known, because Manningham is an area of poor housing and overcrowding, and in these circumstances fires, whether accidental or otherwise, are not uncommon.

The Bradford fires were, it seems, the work of two young Pakistani boys of about the age of twelve and fourteen; they show how even the relatively immature can cause not only havoc and serious damage to property but also loss of life. This case fits in well with the views of Macht and Mach who have studied the 'fire-setting syndrome'. They point out that fire-raising is a 'highly complex and determined piece of behaviour and not simply the product of the breakdown of an impulse.' The Pakistani boys clearly had a highly developed system. They had to buy the various materials and then select how and where they would place them. Apparently the two formed the type of active/passive pair mentioned.

We have already noted that the fire-setting children described by Lauretta Bender had minor degrees of brain disturbances leading to learning difficulties. To support this there is the opinion of Vandersall and Wiener, who indicate that child fire-setters experience 'significant failure at school and indeed failure in many aspects of life'. They found that such children performed less well than their intelligence quotients would have suggested, so at an early age we already find an under-achievement pattern. Interestingly enough this is often prolonged into adult life; part of the 'hero' fire-setter's inner motivation is the need for appreciation and recognition, stemming, in Lauretta Bender's view, from his deprivation of the love and security of home life.

Some of the children studied were in institutions, but even when they were living at home there were usually some serious problems – an unsympathetic step-parent, children from another marriage, an invalid or absent father, or the presence of mother's boy-friends. Racial and religious backgrounds seemed to be of little importance, but one common factor was that even at a very young age fantasies of fire and fire-setting were the rule, the first incident quite often being precipitated by some new and stressful situation in the home. The deprivation of love and affection is a common factor throughout the whole range of conduct disorders in childhood, but in Lauretta Bender's opinion the fire-setting group were particularly deprived of these psychological necessities and in some there was physical deprivation also in regard to food and clothing. Because the parents were unable to provide the children's physical or psychological needs the fantasies became even more pronounced in their aggressiveness and destructiveness, with the parents often being used as a target.

The chaotic nature of the family relationships of young fire-setters is brought out by one case quoted by Vandersall and Wiener. This was of a quiet, affable, nine-year-old boy who openly discussed his struggles with 'evil thoughts' concerning fire-raising: 'When he did set a matchbook on fire his mother saw this as the culmination of defiant behaviour that could not be controlled by her and she sought psychiatric hospitalisation for him.'

This was not in fact carried out and he was supervised as an outpatient for a period. The boy had been reared by several mother substitutes because of the mother's own inability due to her own problems. The boy's care was naturally inconsistent, nor was there a father around to control the boy; as in so many other cases it was the home that was the site chosen for the destructive act. The fire-setting took place soon after the boy and his mother had been re-united and were living with friends. This was in fact the first occasion that he had lived with his mother for five years, and throughout that period he had been given a relatively stable home by his grandmother. It came out in interviews with the boy that 'the move from grandmother to mother had aroused not only a sense of loss, rage and depression, with wishes both for restitution and revenge, but also placed him in a much more intimate relationship with his young mother.' Vandersall and Wiener conclude that 'although he did not elaborate fantasies about fire, it clearly could represent an expression of both destructive and libidinal impulses.'

The younger children's fantasies about fire and their aggressive feelings towards their parents make them extremely anxious, but on the other hand adolescents have learnt to cope with these feelings and are almost guiltless. For the younger ones the resolution can be achieved by magic, burning effigies of father or mother; only later is this attempted in reality. Of course this magical view of fire is not confined to children, it spills over into adult life and is present in the psyche of those who burn themselves to death. In these acts burning the self is considered as destroying the enemy. The deprived child uses magic in a slightly different way. It allows him to grow in power and approach the strength of his parents. Fire is seized upon as a powerful primitive and potent weapon for obtaining results and discharging anxiety created by destructive fantasies.

Sexual aspects
The sexual aspects of fire-setting are often stressed in relation to adolescents, for they have to resolve both sexual and social conflicts with the opposite sex as well as with their own. Freud has made observations on the similarity between the penis and the firemen's hose, between urination and putting out fires with water. We have already referred to Freud's comments on the Prometheus legend, when he pointed out that it was both an aggressive male act towards the gods, and also passive – for punishment Prometheus was chained to the rocks and attacked by vultures. In relation to fire-raising these two roles do appear to be important. However, in practice it is the male who uses fire, presumably to make up for his inadequate masculinity. In the upbringing of a child the interaction with both mother and father are important. Traditional psychoanalysts emphasise the mother and, as we have seen, the mothering tends to be of poor 'quality' in the fire-setting group, but it is probably the father who is the more important since he usually dictates the boundary of behaviour which establishes just what can and cannot be done. Erickson in his *Childhood and Society* points out that the child has good mother control at a time when his emotional relationships and experience of social interaction is limited. This is why, in Erickson's view, the child may resort to magic. The other aspect clearly seen is his use of identification and imitation.

Erickson quotes the interesting story of the 'son of a bombadier' who exemplifies some of the forces at work in the world of the child and shows how these forces may temporarily lead to difficulty. He was the five-year-old child of a neighbour. The boy had previously been well-behaved but suddenly he became defiant, revengeful and destructive, fire-setting being his particular outlet. On one occasion he set fire to a pile of wood, and when chastised by his mother said, 'If this was a German city you would have liked me for it.' He was clearly attempting to identify with his father, a vigorous and muscular man in the armed forces. The child first started a series of fires when his father returned to his unit after having spent some leave at home. On that particular occasion the bombadier had been welcomed home as a hero, and it was this that seemed to spark off the fire-setting. However, the child gave up fire-setting, now channelling his aggressive feelings into his activities on his bicycle. At this he was a virtuoso. He would race towards smaller children and frighten them, and this seemed to satisfy his aggressive urges. No further fires occurred. One can see this type of situation also in adult life; when the individual is deprived of identity, he becomes frustrated and unable to establish a particular role, and this can lead, as Erickson says, 'to murder'. The urge to set fires may well be another, less dangerous equivalent to resolving conflicts.

In summary, fire-setting in childhood can be experimental or part of a behaviour disturbance. However, by adolescence it has already taken on the pattern seen in adult life; it is a complex act carried out without remorse by boys rather than girls and often without any clear motivation. We must now examine some of the characteristics of fire-setting in adults.

Chapter Ten

The Fire-Bugs

> I hear the alarm at dead of night,
> I hear the bells – shouts
> I pass the crowd – I run.
> The sight of flames maddens me with pleasure.
> <div align="right">Walt Whitman, <i>A Song of Joy</i></div>

Looking through the literature on fire-setting from the point of view of the interested psychiatrist, I came upon the above quotation from Walt Whitman, which struck me as pinpointing exactly the feelings of someone who sets fires for the pleasure of seeing the fire brigade arrive, together with all the associated sounds and excitement. Was the poem just an example of the insight that the poet has into many aspects of life, or was it part of his own unconscious desires? Reading what his biographers had to say about him, I found to my surprise that when he was the editor of a new paper in Brooklyn called the *Weekly Freeman*, the office and equipment were destroyed by fire the day after the first issue in 1848. Was this a coincidence? I read Whitman's poetry once more and found that there was a considerable interest in fire, both in reality and as symbolising passion. Another aspect of his life was his concern for the troops during the American Civil War. His work was dressing wounds and in fact a considerable amount of his poetry is devoted to this part of his life. I found also that one of his great interests was to visit in hospital firemen who had been injured in the course of their duties. The other feature of Whitman was his homosexuality. How can we put all this together? We could postulate that he was lacking in 'maleness' and that he made up for this lack by identifying with firemen and soldiers, two groups noted for their vigour and masculinity. No one will ever know if he needed to use fire to satisfy some insatiable inner desire for power and actually burnt down the offices of the *Weekly Freeman*.

There have been many others who have been interested in fires, firemen and putting out blazes. Both George Washington and Benjamin Franklin were regularly enrolled as volunteer firemen. I do not suppose for one moment that they were fire-setters, but we shall see later that there are well-documented examples of firemen whose interest extended not merely to putting out fires but to starting them as well, thus enabling themselves to go through the whole exciting process from beginning to end.

Another famous 'pyromaniac' was the Prince of Wales. He is described by Ronald Pearsall in *The Worm in the Bud: the World of Victorian Sexuality*; when the music hall, the *El Dorado*, was burnt down in 1869 the Prince of Wales, 'always eager to go to fires', was at the scene in the guise of a fire officer, riding on the fire engine. But then many people show and have shown great interest in fires, and in the men who extinguish them, without going on actually to start blazes themselves.

'Fire-bugs' are defined by Lewis and Yarnell, who wrote a detailed study entitled *Pathological Fire-setting*, as 'offenders who said they set their fires for no practical reasons and received no material help for the act. Their only motive was to obtain some sort of sensual satisfaction.' Lewis and Yarnell studied a large number of fire-setters in the United States collected over many years. There were in all 1,145 men and 201 women. In this major work, a highly detailed description of all types of fire-setters, they classified 688 male adult fire-setters as fire-bugs. In other words more than half the men who set fires deliberately were without any real motivation and they carried out this act repeatedly to satisfy some inner desire.

The fire-bugs themselves gave various explanations for their satisfaction; 201 said they wished to help the firemen to be heroes, or to enjoy the destruction of property. Most of the remainder merely said that they had an irresistible desire within themselves to set fires. There seemed to be other aspects as well, such as the desire to see fire engines, while some actually got sexual satisfaction from watching the conflagration they had kindled. This enjoyment of fire was important, and there was denial of any kind of a revenge motive, though sometimes the older fire-raisers thought that the world was against them and expressed other paranoid ideas as well.

Fire-setting sprees
The fire-bug goes out on his own for his fire-setting sprees, which are usually nocturnal. He feels his actions are under no control; it is almost as though he is an automaton, or perhaps a puppet under someone else's influence. The fire-bug thinks of himself as a kind of Jekyll and Hyde, and that the blazes are the result of 'evil' taking over. When confronted with what he has done he realises that it is senseless and unreal, but he insists that it is quite impossible for him to control his behaviour.

The fires are not planned, in contrast to those started for profit. There the arsonist penetrates into the heart of the building and starts the flames with a timing device so that there is a delay allowing him to escape. The fire-bug kindles his fires in hallways, staircases and passages, places to which the general public have access. Like the arsonist, he moves speedily, but he does not prepare in advance, often using the materials at hand, the wastepaper lying near a rubbish chute, or a brimming dustbin. He thrusts a match into it and rushes off, more likely than not to start another blaze. As many as ten or twenty may be kindled within an hour in the same neighbourhood.

A fire-bug has described the mounting tension before one of his outbreaks; when he is in the grip of an attack he is restless and on edge; he has a headache, palpitations and ringing in the ears, and there is a feeling of complete unreality. But to the casual observer he appears to be merely a man in a hurry, and indeed even if in the middle of one of these attacks the fire-bug is challenged and asked what he is up to, he will appear perfectly normal. As a result he is often allowed to go on his dangerous way. Some stay at the scene of their activities until the fire brigades arrive. The fire-bug can throw a whole neighbourhood into confusion, and this appears to be part of the enjoyment; when it is all over he retires to bed. Some fire-bugs even report that they are quite unable to do this until they see that their fires are under control.

The fact that the fire-bug can go home peacefully to rest after such a spree is reminiscent of the calm which follows sexual excitement and satisfaction. Is there any reason to believe that this is a true parallel? Such a view would imply that the fire-bug was unable to obtain sexual satisfaction in the usual way. Certainly there is evidence to support this idea, because only a third are married, compared to over two-thirds of the population of the same age range. Further, even the married ones often have an unsatisfactory relationship ending in separation and divorce. Other characteristics of the fire-bug include his poor physical qualities; he may for instance be disfigured or of low intelligence.

Overall, poor adjustment to life in general is a feature of the fire-bug; his life will often show repeated changes of job, an inability to form stable relationships, and a high alcoholic intake. This last factor may be important, since alcohol tends to decrease inhibition, paralysing the conscience and letting animal feelings have free range.

The world of the fire-bug is concerned only with his pleasure in fire-setting, which has replaced all the usual and varied aspects of life that please well adjusted individuals. This emphasises how difficult any form of treatment would be for the fire-bug. It is similar to the chronic alcoholic whose life is alcohol and without it there is nothing.

The 'motives' of the fire-bug
A fire can be a thrilling spectacle. It is a great drama in which the firemen come into their own and people realise their function as protectors of the community at large. Lewis and Yarnell feel that exhibitionism is an important factor in the makeup of the fire-bug: 'They secretly stage a drama of their own, in which they are author, stage designer and the leading actor.' The fire-bug can be seen as a 'drop-out' from society who nevertheless has the ability to hold the community to ransom, to confound the experts and render the authorities helpless. Not only this, the fire-bug can participate at every level; he can play detective, give first-aid treatment, assist the firemen and finally, if he is arrested, he will be photographed or televised and his picture transmitted nationwide.

There are parallels here with 'exhibitionism', a sexual deviation in which men expose their genitals in public, usually to women. It is an act repeated often, perhaps many times a day, and then not for days or weeks. Another analogy between the sexual pervert and the fire-bug is the 'glorification' which follows his arrest and confession. The sadistic murderer may give himself up, thus enjoying 'the other half' of his perverted desires, namely the masochistic one. A few years ago a brutal murderer of two teen-agers appeared in the offices of the paper with what was then the largest circulation in Britain. There he made his confession; how better to satisfy the masochistic side of his personality than by being featured in the maximum number of copies of a newspaper?

Firemen as fire-setters
The fire brigade is clearly an important feature of industrialised and urban communities, which suffer most from fire damage. Of course the fire brigade has other activities as well as putting out fires; there is fire prevention, there is work in rescuing would-be suicides, helping when houses are flooded and, more important, when disasters occur with trains and planes. But their main duty remains to protect us from the very serious risk in our cities of fire getting out of control.

It is clear that we owe a great deal to the fire brigade and in the larger towns and cities it is highly organised and relatively well paid. On the other hand it is a difficult job in that there are long periods of inactivity followed by rather briefer periods of overwhelming activity. In the periods of waiting there are the military-like duties of cleaning and polishing and the maintenance of equipment, with endless training to maintain physical fitness, not a way of life which appeals to many people.

In smaller towns the fire brigade relies on volunteers, and here 'undesirables' may well be recruited, as Lewis and Yarnell point out: 'It is more likely in a small community that someone would desire to appear as a hero.' Obviously the fire-bug may be attracted to the brigade, since at least he will have the excitement of all the fires in the neighbourhood which he attends. The trouble is that he can extend his pleasure by starting fires himself, and this has occurred from time to time in many parts of the world. Lewis and Yarnell had, in their group of over a thousand fire-setters, a small number of firemen who started blazes on their own and others who worked in groups. The recent group of young French firemen, reported in *The Guardian* of 24 August 1972, who started a series of blazes were clearly highly delinquent; not content with their fire-raising activities they actually attempted to derail the Paris-Brussels express by placing an obstruction on the line.

Case history of a fire-setting fireman
'Len' was a twenty-nine-year-old fireman working in a midland town, as his father had and his grandfather before him. His father however had developed heart disease and taken to drink while still quite a young man. He and Len had always quarrelled and only occasionally did they make any real contact with each other. Len's mother was a somewhat shadowy figure; she had raised six children, and was always 'tired', as she put it. Len was a sickly child who was teased at school because he was a 'weakling'; he used also to wet the bed until he left school. Though he was not simple, his academic record at school was not distinguished, and the menial job he took delivering groceries seemed appropriate to his abilities. However he changed his job frequently. When he joined the fire brigade he seemed to settle down, and soon he married a local girl whom he had known for many years. She became pregnant, but much to his disappointment had a daughter. Though there was no particular problem with the pregnancy or delivery, and the child was not particularly difficult to look after, Len's wife remained a semi-invalid. She was constantly complaining of 'period pains' and was never again 'warm' to him sexually, though even before the birth of the daughter they had not been especially excited by each other. At this stage there was a fire in the neighbourhood, but no cause for it was found. It was in a tool shed on an area of ground used for allotments. The next thing that happened was that Len's mother died. Her passing upset him considerably and he seemed depressed for weeks, staying at home during his off-duty periods from the brigade rather than meeting his friends for a drink. In this period there were two further fires in the neighbourhood, both in unoccupied buildings, one in a shed and the other in the outbuildings of a farm. Then Len seemed to get over his depression and there was some improvement in his relationship with his wife. She became pregnant again and

produced another daughter. This clearly upset Len once more, who became quite depressed and uninterested in life in general. He showed no enthusiasm for the new baby, though his wife if anything took more interest and became much more lively and capable.

This improvement in the wife's health and well-being was not reflected in Len's state of mind. Soon a new series of fires started in the locality. They were all in the early evening of the same summer, which had been long and hot. The fires were separated at first by a week or so and then by only a day or two, and they were always in unoccupied premises of one sort or another. Later it turned out that Len, who had a small workshop at the bottom of his garden, would say to his wife that he was going to mend his bicycle and then would climb over the fence at the bottom of the garden to roam in the area beyond. It was only when the blazes were plotted out on a map that suspicion fell on Len. He was always a very good member of the brigade, who reported to the station with great alacrity when the alarm was raised. The fires were now regarded with concern and a more thorough investigation was mounted. In one incident it was noted that a smouldering blanket seemed to be the seat of the fire. This, it turned out, had belonged to his elder daughter and it was the main piece of evidence used to convict Len. The fires he set seemed to occur when he was in a depressed state. The causes however of the depressive episodes were rather trivial, the main aspect of the whole affair resting on his tenuous personality adjustment. Len was sentenced to a long term of imprisonment, which seemed inappropriate because it could not affect his overall personality adjustment. Its only purpose was to prevent sheds being burnt down in a small midland town.

False alarms
There are certain people who hang around fire stations because they are interested in the activities of the brigade. Some of these people have failed to reach the requirements for joining the fire service, and it is often these who get involved in fire-setting activities. Among them also there is a certain type of person, less dangerous than the fire-setters, who merely send out false fire alarms in order to see the brigade arrive. These are usually children, or young people of limited intelligence, and the prank is

often played by groups. Unhappily, even some of these false alarms can end in disaster, when the fire brigade is involved in an accident, or the brigade is called away just before a genuine fire alarm is received. In the report of Her Majesty's Chief Inspector of Fire Services for the year 1970 it is noted that there had been an increase in the number of false alarms; the report points out that the fines imposed, although the maximum is £50, actually averaged only just over £7.

Vagrant fire-setters
All large cities have a decaying area in which are condemned houses, demolition sites, and empty spaces where the vagrant population collects to drink its cheap wine and cider. In the winter these people have a problem, for they must find warmth during the night and sometimes in a drunken and befuddled state they light a fire in a half-demolished building to warm themselves. Often it gets out of control and results in a serious conflagration. This type of fire occurs commonly in the East End of London, the area in which my own hospital is situated. But this kind of thing is not confined to London: it is well known all over the world. These vagrants, without home and family, have often dropped out from society many years ago and the fires they cause are largely accidental. However, this is not true of all of them, for there are well-documented instances of tramps setting fires deliberately. Such tramps have a general grudge against society, roaming the country, hitch-hiking and – especially in the United States – riding illicitly on goods trains. They are people of low intelligence and may be mentally subnormal; in some instances they show definite psychiatric disorders, such as schizophrenia. Their desire to wander, coupled with their interest in fire, is a fatal combination, since before any investigation can be underway the tramp is off to his next location.

The American investigators Lewis and Yarnell have made some interesting observations. They report the case of a thirty-nine-year-old man, of low intelligence, who was finally brought to justice. It was found that he had touched off a whole series of fires between 1933 and 1942, involving box cars, freight-carrying rail wagons, barns, bridges, shacks and a lumber company. The damage done by this type of wandering fire-setter can be immense.

Sometimes these wandering fire-bugs 'hunt in pairs'. There is another notorious American case, that of Stine and Walzer, two fire-bugs who started freight train fires. In the course of a little over a year they had a total of a hundred blazes to their credit spread over practically every state in the country.

These vagrant fire-setters have an even poorer adjustment to life than other fire-bugs. They have often had an institutional upbringing and their only gratification appears to be through their primitive interest in fire. To the psychiatrists they represent an almost impossible case for treatment.

Women fire-setters

Women turn to fire for the solution of their problems less often than do men, at least in the twentieth century. Strangely, this is somewhat at variance with the earlier Continental writers (for example Krafft-Ebing) who comment on the frequency with which women set fires. Lewis and Yarnell found that only 14.8 per cent of their fire-setters were women, and there was also a significant age difference: the peak incidence for men was in the age group sixteen to twenty-five, while in women there was no particular age when fire-raising occurred. Women, however, can be dangerous and dramatic in their fire-setting activities, though in many respects their motives are often the same as their male counterparts. They want to attract attention and to create excitement, but most of all they want to wreak revenge, either against husbands or more usually against lovers.

One difference between male and female fire-setters is that women have little interest in the arrival of the firemen and the fire engines. If they do want to add some kind of heroic aspect to the scene it is usually in the form of the heroic victim. The female fire-setter does not seem to show a grudge against the world at large and the danger she represents to the community is small. Generally, women start fires in their own home or in other property belonging to them, sometimes dressing for the occasion in fantastic costumes so that they can be in the centre of the stage. They are most destructive to their own and their families' possessions, and if they extend beyond such bounds it is usually to a neighbour's property or possibly their employer's or ex-employer's. Quite a high proportion of fire-setting in women is associated with suicide; sometimes these women see their children as extensions of themselves and then, in a self-destructive phase, they may well set fire to their children's beds.

One might have expected adolescence to produce a peak in the incidence of fire-setting in women, since it is a time when emotional stress is generally high. Perhaps the enlightened attitude to the significance of menstruation nowadays, and the removal of the anxiety about it, has led to a decrease in the amount of female fire-setting. The nineteenth-century literature concentrated on this relationship between anxieties about menstruation and fire-setting activities. Today, female adolescent fire-setting seems more generally to occur against a background of promiscuity and petty stealing. Fire-setting also seems to be a need in psychotic women more than in psychotic men, although overall the female numbers are much smaller.

'Love me or I'll burn you down.' This is the way a fourteen year old, physically well-developed girl of high intelligence felt about her twenty-two year old choirmaster. It is a case of an adolescent female fire-setter quoted by Lewis and Yarnell. The trouble was he gave her the 'brush off', and she was asked to leave the choir. Thus this is a typical case, in that general emotional problems, rather than those relating to menstruation, loom large. The adolescent jilted by the choirmaster 'took the torch she was carrying for him and touched off a fire'. In her purse was found the following note:

> This bright morning I came upon a plan which no doubt proves successful, but not very brilliant. I went to the church and threw all the vestments on the floor. I had no idea of starting a fire then. Next day I went back. I got some cardboard and put it next to the vestments and lit it with a match. I waited until it was burning. Then I went to a girl-friend's. The next day the papers said only a little damage was done. Some day – it may take years – I will start another fire and destroy the whole structure. This I vow.

When she was taken to court it was ruled that it was a childish and impulsive act and she was placed on probation. We do not know the end of the story, but we do know that the girl was constantly running away and having to be returned home. This was obviously a case of adolescent emotional problems for which she herself was unable to find any solution. Such girls, with increasing age and psychiatric help, often settle quite speedily into the role of a happily married mother.

Older women may set fires in the homes of their parents. The interesting aspect of this is the apparently trivial nature of the actual precipitant for the incident, though the background is often one of great conflict, particularly between mother and daughter. But the blaze may also be kindled in order to force parents to buy new clothes and furniture. Indeed one might consider such an act as more characteristic of mental derangement, but it occurs in otherwise rational individuals. The trivial explanation clearly places this kind of act in the 'motiveless' group.

An intriguing story in this connection was reported by the *Sunday Express* on 14 March 1971:

WHY LONELY WIFE SET FIRE TO THREE FARMS

When three farms were destroyed by fire within a week near a village in the Austrian Tyrol, police were sure it was the work of a fire-bug.

But detectives could think of no logical motive. It could not have been gain because one of the farms had not been insured at all, a second only partly. Was it the work of a madman? In that case, nobody could predict where he would strike again.

The answer came when police arrested a twenty-seven-year-old farmer's wife and charged her with setting the farms alight.

Sobbing bitterly, the beautiful mother of two daughters aged six and seven told police: 'I am responsible for the fires. But it would never have happened if my husband had not started to neglect me.

'Until the autumn of last year we were very happy together. Then he started to spend his free time in the village pub with his pals, drinking and playing cards, and not coming home until the early hours of the morning.

'I often begged him to stay at home, but it was in vain. Then I had an inspiration. If he was afraid that something might happen to our farm in his absence, he would stay at home and I would have him for myself again.

'I was right. After the three farms had burned down he stayed at home every evening watching out in case the fire-bug visited us. But now you have spoiled it all – he is sure to go back to his bad habits!'

Unfaithfulness is, as in the above instance, not an uncommon reason given for setting fires. On occasions it is aimed at a female rival but this is more unusual; directing the aggression towards the property of unfaithful husbands is much more characteristic.

The abused servant girl setting fire to the home of her master and mistress seems to have a ring of a previous century and indeed it was the kind of thing reported by the nineteenth-century European psychiatrists. The maid might set fire to the master's bed, either because she was jealous of her mistress or because the master had made 'a pass at her' and she felt affronted. Some of these fires seem to be associated with definite sexual arousal and the act may lead in some instances to a feeling akin to an orgasm.

Women with psychotic illness may, as do men, set fires in response to their hallucinatory voices or in relation to their delusional systems. To this we shall return in the next two chapters. But they rarely seem to set fires in pairs, nor do they often kindle blazes to cheat insurance companies. Instances are recorded of female fire-setters who set more than one blaze, but these constitute only about 10 per cent of the total; even then each fire is usually started for a specific reason and they may be widely separated in time. This is quite different from the activities of the male fire-bug who, as we have seen, sets repeated fires, usually within a short space of time.

Why do women turn to fire-setting? Why do they appear to be different in some respects from their male counterparts? In general women fire-setters are not different from other delinquent women. They suffer from problems of immaturity and from mixed feelings in relation to their sexual role, a feature discussed in detail recently by Tennett and colleagues. Female arsonists rebel against the man-made code while still retaining their female intuitions and interests and wanting to be loved by their menfolk. They tend to get on badly with their mothers and to idolise their fathers. The symbolism of fire takes on two main forms for these women. First of all, in its sadistic aggressive form, it represents the ability of the woman to destroy her mate. Next, in the much more important masochistic form, it represents a sexual attack, destroying the woman completely but in a glorious and all-embracing manner. This view of the self-destructive aspect of fire is much more prevalent in women than men, while the intense excitement that surrounds the physical strength and energy of the fire engines and the firemen which appeals to the male fire-setter is something not found in women.

Chapter Eleven

Jonathan Martin, Incendiarist of York Minster

There are some individuals whose claim to fame, or perhaps more correctly to notoriety, rests on a single act, an act of rebellion against society. Such is Guy Fawkes, or the man who in ancient Greece destroyed the theatre at Epidaurus by fire. Jonathan Martin is another such person, known only for his arson attack on York Minster. He is an example of a type of fire-setter that we have not encountered so far.

Jonathan Martin, born in Hexham in 1782, was the third of the five children of Fennick and Isabella Martin. His younger brother John, born in 1789, was an artist in whom there is considerable current interest; his older two brothers William (1772) and Richard (date of birth unknown) both had something to their credit. William was called 'the philosophical conqueror of all nations' and Richard wrote poetry.

Jonathan is particularly interesting from our point of view because the documentation of his life and thoughts is fairly complete. It is based essentially on his autobiography and amplified by the researches of Thomas Balston who wrote *The Life of Jonathan Martin – Incendiarist of York Minster*.

Jonathan apparently was slow at learning to talk, and was also apt to wander by himself even in the middle of the night. On one occasion he went to a nearby lead mine to watch the men working; they saw him standing there and at first thought he was a ghost. He ran away but was caught and thereafter his nocturnal prowls were forbidden. Another early memory, according to Balston, is of 'his mother's instructing him that there is

God to serve and hell to shun, and that all liars and swearers are burnt in hell with the devil and his angels.'

There is also a curious story of some disturbance in the night and of a child being thrown downstairs. Apparently a neighbour confessed to the act and was 'committed to prison, found guilty and died a miserable death'. The real story behind this event is unknown, but it would appear that Jonathan came from what in modern parlance would be called a 'disturbed background'. Another interpretation might be that he had some terrifying nightmares as a young child. It was at this early period in his life that he slept, at least on one night, in his mother's bed when his father was away from home, a practice which is considered by child psychiatrists to be highly disturbing for young children.

In 1804, when Jonathan was twenty-one, he joined the navy. In his book he tells of the colourful life that then followed, including 'the deliverances the Lord hath wrought for me'. On three occasions he was almost drowned.

A little later he found himself on board a ship called the *Hercules*, which was engaged in the Peninsular War. It was at this time that, again according to Balston, the following incident happened. 'From Lisbon they sailed to Cadiz and Jonathan was in mid-gunners' crew. On the voyage the gunner's yeoman who was in charge of all the powder and stores, shot himself through the head in the storeroom, where there were five hundred barrels of gunpowder.' Of course this caused consternation among the ship's company, who thought that the boat might explode at any moment. Of the six hundred on board some leapt overboard, others took to the boats, but 'Jonathan, with four of his shipmates, rushed into the storeroom, which was full of smoke, soon extinguished the little fire produced by the wadding of the pistol and then discovered the unfortunate man lying bleeding with his brains strewn all over the floor.'

At Cadiz, following many narrow escapes from shore batteries Jonathan began to feel a special relationship with God. He also began to see himself as a sinner and he decided to return home to join the 'people of God', the description at that time of the Wesleyan Methodists. To do this he had to leave the navy; while all hands were on deck, having been 'piped to breakfast', he dropped unseen into a boat that he had put under the bows and got ashore. He stayed in the safety of the waterman's house until the ship set sail.

Balston was able to find an account by a shipmate of Jonathan in Leeman Thomas Rede's *York Castle in the 19th Century*. Rede remembered Martin as a playful sort of person but one who nevertheless had the nickname of 'Parson's axe'. Occasionally he was sulky and idle and though he did not pray much he was inclined to argue on religious subjects and 'said he had a light that we had not, and that he had "meetings" in his dreams'. Apparently he was 'fond of viewing and conversing about the celestial bodies but had a dread of anyone pointing to a star, a thing, of course, of common occurrence at sea, and would not believe that there were other worlds, and indeed grew angry at such an assertion.' He also talked of a book being shot from his hands when he was at Cadiz and this Jonathan apparently considered to be 'a warning from heaven'.

From another account of a Greenwich pensioner we have the view that Jonathan was subject to fits of melancholy and then he would 'talk of dying and of a future state'. At other times apparently he was overactive, larking about in a dangerous fashion, so much so that he once fell from the rigging and was injured.

Both this pensioner and another corroborated the story of the fire in the storeroom and neither apparently thought that Jonathan was responsible for this.

Jonathan's movements until 1818, that is in the eight years after he left the navy in 1810, cannot be ascertained. However, he married and had a son in 1814. He also began to ruminate on his life at sea with its 'profane talk and licentious songs', and pondered on whether he had committed terrible sins and whether he was worthy to take 'the sacrament'. Sometimes he went to church and at other times he went to the chapel, finally deciding that the chapel was right for him. Apparently, 'one Sunday he took the sacrament at Stockton Church and when the service ended at 1.30 p.m. he then ran the four miles to Yarm and reached the [Methodist] love-feast there, which began at 2 o'clock.' Before the first prayer was ended the next night he had his reward. He attended the full prayer-meeting at Norton, and while the class leader was giving out the hymn, his conversion was achieved. 'The spirit of God came down on me in such abundance that I was ready to leap over the table for joy. We continued in prayer till 10 o'clock, till we had all prayed twice. While I was engaged in prayer I was powerfully assisted by the spirit; and there was not a dry eye in the house.'

There were still worries in his mind about leaving the established Church and it seemed at this time that he was somewhat deranged. One Sunday morning, for instance, he went to church early, apparently in obedience to a voice. He walked round the church seven times, and on the seventh time the clerk arrived and he was allowed inside. He visited the vaults, then climbed into the pulpit, shut the door and lay down. As the Sunday morning congregation assembled Jonathan still lay there, until he was pulled down from the pulpit by the clerk and put in a pew. When he returned home he was greeted by his wife who had heard of his doings and was relieved to know that the village constable had refused to become involved.

Jonathan was constantly berating the Church of England clergy for their laxity of discipline and the laity for their scandalous lives, so much so that even the Methodists were alarmed by his conduct and wondered if they should expel him from their society. He was already finding difficulties in getting a job after having been dismissed as a result of his peculiar activities. Now he began to write letters to the clergy, attaching them to the church door; he also attempted to address a congregation at Bishop Auckland, but as soon as he began to speak he was arrested as a vagabond. Later, when he heard that Dr Edward Legg, the Bishop of Oxford, was to hold a confirmation at Stockton he decided to test the bishop by pretending to shoot him. For this purpose he borrowed a pistol from his brother. It had an old broken barrel tied together with string and no one would have dared to shoot it as the weapon would certainly have exploded. He confided to his wife that he was going to shoot the bishop, but the next morning the pistol had disappeared; presumably she had got rid of it.

At the confirmation Jonathan commented on the wonderful size of the bishop, which he considered was due to excessive drinking. But his wife was becoming increasingly alarmed by his threats and presumably she talked to the vicar; as a result, on the next day Jonathan was arrested as a dangerous lunatic. He was ordered by the justices to be confined to a madhouse for life and the next day was removed to the lunatic asylum at West Auckland.

Later, with the help of his friends, he was transferred to the asylum at Gateshead. There he was much happier and more content though still anxious to be freed. A Mr Nicholson, who was in charge of the institution at that time, reported at Jonathan's trial eleven years later that he was always composed and tranquil. On 27 December 1888 Nicholson sold the asylum to a Mr and Mrs Orton. After some initial difficulties he was allowed the freedom that had been his with the Nicholsons and was able to work in the garden and the bakery. When Mrs Orton subsequently gave evidence at Jonathan's trial she brought a picture painted by him dating from this period. It was reported to be an extraordinary mixture of talent, frenzy and wildness. It described the Christian's progress through the various steps of his religious experience, his heart pierced by the sword of the spirit and set on fire by divine love. Also in the picture was the hymn of Charles Wesley.

> See how great a flame aspires,
> Kindled by a spark of grey;
> Jesus' love the nations fires,
> Sets the kingdoms in a blaze.
>
> To bring fire on earth he came;
> Kindled in his heart it is;
> Oh, that all might catch the flame,
> All partake the glorious bliss.

This, as we shall see, turned out to be disastrously apt in relation to his attack on York Minster.

The next part of Jonathan's life turned out badly. He decided to walk to Ravensworth Castle and in his possession he had the keys of the storeroom of the Gateshead asylum. He was 'recaptured' and restrained with irons riveted on his legs and in a room with windows double-barred. Later, he was again allowed the freedom of the garden, but he was still fettered. One day he managed to file away the rivets of his fetters with a piece of stone, and in this way he was able to make an escape by climbing through the roof, an event pictured in his life and captioned

'Jonathan Martin's providential escape from the asylum house'.

After his break from the asylum, Jonathan wandered the country seeking help from friends and relations. At length, he decided to go back to Norton, where he had lived before. 'To his surprise at the entrance to the village he met the magistrate who had committed him to the asylum. The magistrate questioned him as to how long he had been out of prison, and how he got away, and where he was going.' The magistrate, suggesting that he leave or he would be recommitted, gave him a shilling.

He returned to Edward Kell, a distant relative, who had helped him soon after his escape from the asylum; there he stayed, 'meditating on the goodness of God and enjoying sweet communion with his saviour'.

But Jonathan could not keep away from Norton and soon he was back again. Mr Page, his previous employer, 'ordered him to start work as soon as he found it convenient, and the magistrate agreed'. However, the vicar, on his return after a temporary absence, requested that Jonathan be moved elsewhere; he went to Darlington.

In 1822, soon after his arrival in Darlington, Jonathan found a tanner and Methodist called George Middleton who employed him and who allowed him to attend the 'love-feasts' of the Methodists in the neighbourhood. Jonathan soon began to convert his workmates, telling them of their wicked lives and admonishing them to repent. He suggested to them that they should fall down upon their knees and cry for mercy rather than go to the alehouse. He himself rose at 5 o'clock in the morning to read the Bible.

But still Jonathan's troubles were not over. One day he asked his master, George Middleton, to reprove his sons for hunting, which he did, but of course this annoyed Middleton's sons, who began to persecute Jonathan, 'filling his clogs with dirt, throwing his boots and clogs into the pit, and heaving wet skins at him'. One son continued to hunt, and Jonathan prayed that God would take away the pony: 'Thereupon the pony fell ill and would not eat and one day, when it was being led about the yard and young Middleton again raised a shout for the hounds, a sudden confidence rose in Jonathan's heart that the Lord had heard his prayer.' Though he was scoffed at, the pony died four hours later and this nearly ended in Jonathan being burnt as a witch.

The next period, from 1823 onwards, is somewhat vague, partly because Jonathan's autobiography ends abruptly at this point, but he probably stayed in Darlington or its surroundings. However, the next important landmark was his arrival in Lincoln in September 1827.

One episode in the events of his life soon after his arrival is recounted in a single paragraph at the end of the third edition of his Life. In company with a young Methodist, he was holding a meeting near the cathedral. Jonathan prayed at length, 'that God would fill Lincoln Cathedral with converted clergymen and distribute them among all the churches of Great Britain'. He clearly felt that this would disturb the devil in his den. Be that as it may, it certainly upset the landlady of the nearby public house, who with a friend made a violent assault on him and threw him from a window. But Jonathan landed on his feet, surviving unscathed to sum up the incident as follows. 'The devil was conquered; glory be to God.'

He obtained work as a tanner in Lincoln, but he was still unsettled, wandering all over the north of England hawking his book, preaching, and attaching copies of his letters to church doors. Returning again to Lincoln, Jonathan was readmitted to the Wesleyan Methodist Society. But soon he was on his travels again, and by 28 December 1828 we find him taking lodgings with a Mr William Lawn in York. There he occupied himself in attending meetings of the Methodists and ranters; on Sunday evenings he went to York Minster, and in the evenings he read the Bible and sang hymns, talking constantly of his dreams and particularly that he was the son of Napoleon Bonaparte. In fact he saw himself at the head of countless armies whose mission would be to capture the country. Mrs Lawn did not, however, regard him as insane and Mr Lawn considered him a 'very religious man'.

On 27 December, the day after his arrival in York, Jonathan issued the first of his five warnings to the clergy of York. In the first he advised the clergymen to flee from the wrath to come and repent of their sins. The letter was stuck on one of the spikes of the gate at the south aisle of the Minster. Another, written on 5 January, was in a similar vein: 'Your great churches and minsters will come grappling down.' This note too was attached to the gate at the south aisle. The next day another letter was written: 'You blind clergymen, you who are pricking for your bottles of wine and roast beef and plum puddings and your loaves and fishes.'

The fourth, on 16 January, read as follows: 'You blind hypocrites, you serpents and vipers of hell, you wine bibers and beef eaters, whose eyes stand out of the fatness.' On this occasion the letter was delivered in a different fashion. A missile made up of the letter and a copy of the *Life of Jonathan Martin* was wrapped round a stone and thrown through one of the south aisle windows of the Minster. All traces of the fifth letter have been lost.

Strangely, only one of the letters was sent on to anyone in authority; presumably no one really took any notice. Jonathan himself was disappointed at not receiving an answer in spite of the fact that all of them carried his signature and his address. He waited anxiously for instructions from above. As he said at the trial:

> At length I dreamt that there was a wonderful thick cloud came from the heavens, and rested upon the Cathedral; and then it rolled over, and rested upon the lodgings where I slept. When I found it come, I awoke and wondered what it was and what it meant, for I expected the house to be destroyed. The house shook wonderfully, so that I was awakened out of my sleep. I was surprised and I was astonished. I prayed to the Lord, and asked the Lord what it meant; and I was told by the Lord that I was to destroy the Cathedral, on account of the clergy going to play balls, playing at cards, and drinking. I thought I heard a voice inwardly speaking, informing me that the Lord showed me the vision of the dark to point out the propriety of setting the Cathedral on fire, and that I should make it shake and tremble. I had that so impressed on my mind that I could not rest either day or night. I found the Lord was determined to have me show this people a warning to flee from the wrath to come.

Jonathan's decision had been made. That evening he went to the service at the Minster, commenting that the organ made such a buzzing noise that he said to himself, 'I'll have thee down tonight: thou shalt buzz no more.' At the end of the service he slipped into the north transept and when the building was locked he was alone inside. Jonathan still did not know exactly what he was going to do, but he groped his way into the bell chamber and asked for the Lord's guidance. The Lord said 'Strike a light'. This he did with the Lawns' tinderbox which he had taken without their knowledge, lighting a penny candle. Next he cut down a hundred feet of bell rope, which with his experience at sea he was able to make into a means of escape.

Jonathan pondered on what to burn, finally deciding on prayer books and music books rather than the Bible. He collected these together as well as some cushions, making two piles. He reported later, 'I had had a hard night's work but the Lord helped me.' He scaled the rope ladder he had made and escaped, taking with him velvet curtains, fringes, tassles and a little Bible. As he left he looked back and saw the flames at work.

The fire was not discovered until dawn, when a chorister arriving for early practice noticed smoke coming from a window. He called to one of the masons but when they tried to enter the Minster they were overcome by thick smoke and had to leave.

The problem was now how to put out the blaze. An efficient fire engine could easily have done this but these were not available at the time. Another problem was that the seriousness of the archbishop's throne, as well as other furnishings and the destruction of the Minster now went on apace. Balston reports, 'In the twinkling of an eye the whole organ was wrapped in flames, and, as it blazed, it gave out an unearthly noise which reverberated through the Minster, and alarmed the crowd which had gathered outside.'

Soon streams of molten lead were pouring off the roof and pieces of burning timber falling all around. By the time the clock had struck nine the first part of the roof had fallen in with a tremendous crash; the fire was not under control until the evening. Jonathan's work had completely destroyed the fourteenth-century carved oak roof of the choir, the choir stalls, the pulpit and the archbishop's throne as well as other furnishings and the organ. However, the great east window was undamaged.

The committee of the cathedral clergy realised it was the work of an incendiarist, and it was probably Lawn who was the first to identify Jonathan Martin as the culprit. A poster issued on

5 February 1829 reads: 'Whereas Jonathan Martin stands charged with having on the night of 1 February instant wilfully set fire to York Minster. A reward of £100 will be paid on his being apprehended and lodged in any of His Majesty's gaols.' He was arrested in Hexham and put on trial in York. When he talked of his reasons for starting the fire they were delivered, according to Balston, 'in a tone and manner which indicated that he had no feeling at all in the matter, as if it was a mere everyday occurrence which did not interest him'. It was then that Jonathan recounted his dream of the thick cloud settling over the Minster. During his time in gaol awaiting trial he was sometimes depressed and at other times restless and excited. It was reported by a doctor by the name of Wake that he consumed in a most voracious manner a beef steak pie which his wife had brought him, 'as much as any three men could have eaten'. Another cause for his excitement was a dream he had of a man on top of the gaol wall with an angel assisting him to get down, which presumably Jonathan saw as a sign that he would be delivered from the prison by the intervention of God. At this period he spent much time studying his Bible and hymn book, and read aloud from Isaiah, chapter 64, which contains the verse: 'Our Holy and our beautiful house, where our fathers praised thee, is burned up with fire: and all our pleasant things are laid waste.'

His brothers William and Richard arrived in York to visit him, and Jonathan was exasperated to find that Richard was arranging to prove him insane. His artist brother John did not come to York but it was he who undertook the whole expense of the defence.

On Monday 30 March 1829 the trial of Jonathan Martin began. He was indicted for maliciously and feloniously setting fire to the Cathedral Church of St Peter of York, in reply to which charge Jonathan answered. 'It was not me, my Lord, but my God did it.'

It is important to realise that Jonathan was being tried for his life, as arson in 1829 was a capital offence; hence the importance of any testimony which might prove him insane, thus leading to a moderation of the sentence. In order to establish whether he was mentally deranged at the time of the fire, evidence was called on his previous mental health. The keepers of the Gateshead asylum, the Nicholsons and the Ortons, were thus witnesses, as were many of his friends.

The other issue at Jonathan's trial was the question of prejudice against a minority group. As we have seen with the Great Fire of London, for example, suspicion is always cast on minorities. As Jonathan was a Methodist, and obviously against the established Church, strong feelings of prejudice could have been aroused in both the judge and jury. And the fact that he had set fire to the archbishop's throne and destroyed a magnificent and much-loved building obviously increased considerably the risk of prejudice. However, Methodist witnesses were called who testified that their society had no rooted objection to the prayer book, nor did they believe that the clergy were 'the blind leading the blind', a statement which Jonathan had often made. By assembling such Methodist witnesses the prosecution were able to make sure that the jury were not biased in this respect.

It was quite clear throughout the trial that Jonathan had set fire to the Minster, so the main point for the prosecution was whether or not he had been incapable of distinguishing right from wrong at the time of the act. Doctors were called as witnesses, including Caleb Williams, a surgeon, who was one of the staff of the lunatic asylum at York called 'The Friends' Retreat'. He had visited Jonathan in gaol on eight or nine occasions and, in his view, he was a 'monomaniac', under the influence of dreams when he committed the act. Balston says of Jonathan that 'he was more under the influence of dreams than a sane man would be: when he was excited, his eyes were red, and his pulse full, hard, strong, and quicker than natural.' Dr Baldwin Wake also described Jonathan as a monomaniac.

The jury came to their decision in seven minutes, that Jonathan was guilty but insane. The judge said: 'Your verdict must be that of "not guilty" on the grounds of insanity and he must remain in custody during His Majesty's pleasure.'

Jonathan arrived on 28 April 1829 at the criminal lunatic asylum in St George's Fields, popularly known as Bedlam, the building which now houses the Imperial War Museum. He was there until he died nine years later. Apparently Jonathan remained rational while at Bedlam and rarely spoke of his crime. He maintained a passion for drawing, and his drawings were usually concerned with the clergy. He became excited at times, especially when he was drawing, but also when he was deprived of pencil and paper.

Trying to arrive at a diagnosis on a historical character presents two main difficulties. First, there is the problem that the evidence has already been sifted and maybe is biased; and secondly, the diagnostic label given by the medical men at the time is often archaic.

This applies to the case of Jonathan Martin, as the term 'monomania' is no longer extant in psychiatric parlance. It was Pinel, a French psychiatrist, who first put forward the idea that there was a type of insanity in which judgment was disturbed while the intellect remained intact, and his disciple Esquirol, using the same kind of general concept, had put forward the term 'monomania'. However his ideals were a little different from Pinel's, for he defined it as a partial disturbance of the intelligence, emotions, or will, limited to a single object. As the 'will' was involved, medico-legal difficulties arose when this particular diagnostic label was given.

In 1833 Marc classified fire-setting as *monomanie incendiaire*. He also invented the term for it – pyromania. There was much discussion in both German and French psychiatric circles as to whether the pyromaniac was insane or not. Indeed, some felt that if this diagnosis were made on a particular person then he was not culpable in law and should be incarcerated in an asylum rather than in prison.

Marc considered that all incendiarists were suffering from pyromania: 'Man, the sport of his passions, becomes an incendiary through jealousy and a spirit of revenge. Some want to conceal other crimes, some are motivated by a diabolical love of excitement. Some work to procure liberty, as the insane who set fire to their dwelling house.' He regarded the pyromaniac as in a state comparable to religious exaltation. Further, he elicited from many their desire to be delivered by fire from anxiety, envy, violent rage, and vengeance.

Nowadays how would we diagnose the conditions from which Jonathan suffered? Let us first of all recapitulate the salient features of his life. Jonathan we know was disturbed enough even before the incident at York Minster, to be incarcerated for long stretches in asylums. He seemed to have had periods of excitement and elation while at other times he was despairing and hopeless. In addition, his religiosity was extreme and sprang from visions of God and direct commands, so much so that his actions became bizarre. He also had the notion that people were against him, as well as the belief that he was related to Napoleon. To give these psychiatric terms, he had paranoid and grandiose delusions. This picture suggests perhaps a diagnosis of schizophrenia, but then we have to explain why so many of his contemporaries regarded him as a rational man at least for long periods. The best interpretation would be that he suffered from a manic depressive psychosis, a state in which just that type of alteration in mood and activity which he showed can be repeated time and time again, though in the intervening periods there is no obvious mental disturbance. It was in one of these phases of excitement, during which he heard voices and saw visions – in psychiatric terminology auditory and visual hallucinations – that he set fire to York Minster.

Apart from the internal evidence of his own life, is there anything else to support the diagnosis of manic depressive psychosis? We know nowadays that a small proportion of fire-setters like Jonathan experience exactly these swings of mood, without any apparent external stimulus. Further, this condition is one form of mental disorder in which there is a strong inherited factor. It is therefore of great interest that Richard, Jonathan's son, killed himself three months after his father's death in 1838. Previously the son had experienced a stretch of mental illness in which he thought that he was developing typhus, and that his breath was poisoning his family and turning them black, symptoms characteristic of the most severe forms of depression and rarely seen nowadays. We therefore have some evidence that Jonathan's son also suffered from severe manic depressive moods, which he may well have inherited from his father.

Chapter Twelve
Psychotic Fire-Setters

Psychiatrists generally classify mental illness into three broad categories: neuroses, psychoses and personality disorders. Neuroses are disorders which are important for the individual but which nevertheless are compatible with the continuation of a fairly normal life. Psychoses are serious disturbances, including schizophrenia, characterised by total disorganisation of thought and, often, the hearing of hallucinatory voices, causing severe upset of life in general. Personality disorders are prolonged in their effect, leading to a distorted view of interactions between the individual and the external world. From the point of view of fire-setting, neuroses are of little significance, though at least in young children the act of starting a blaze or indeed the fantasies of its destructive aspects may lead to anxiety. On the other hand, personality disorder is a category which is of the greatest significance, and in particular its sub-category the 'psychopathic personality'. Into this latter group would be placed all those whom I have called fire-bugs. This brings us to the topic of the present chapter, the psychotic group of fire-setters which in the view of Lewis and Yarnell account for just over 10 per cent of all fire-raisers. The range of mental disorder which they display is extremely wide and we shall now examine some of the different types in details.

The manic depressive
Jonathan Martin, discussed in the previous chapter, is an excellent example of one particular type of psychotic fire-setter – the manic depressive. It is in fact one of the commonest psychiatric disorders of the present time. There are, we have seen, inter-

relationships between aggression and depression and between violence and suicide, as the troubles in Northern Ireland reveal. It must be emphasised that although feelings of sadness are within the experience of us all, a depressive illness is much more severe. It is so extreme that bodily functions are upset, appetite is poor, constipation occurs, sleep is disturbed and mental processes and ordinary physical activity may be so slowed down that the person is reduced to sitting and staring into space, unable even to weep. The person experiences feelings of guilt, hopelessness and utter despair which can lead to suicide.

However the obverse of the coin, the manic phase, though rarer, is almost equally alarming. The overactivity begins with rising early in the morning, and continues with endless activity about the house, much of which tends to be purposeless, so that the cleaning and the polishing done by the manic housewife will be pretty slapdash. Shopping sprees occur, with inappropriate purchases of clothes and other more expensive things, and items such as cars may be ordered in pairs rather than singly. High spirits accompany this overactivity, and this may lead to fires started either accidentally or inappropriately. One manic patient of mine set fire to the curtains in her hospital ward because she said she was cold. Another did it as a high-spirited prank and when asked why, replied, 'Wasn't it lucky I was here to raise the alarm?' In this state of elation and excitement the patient will often make the most dour psychiatrist burst into helpless laughter, but such laughter may make the patient angry and aggressive. Indeed, there is frequently in these patients an undercurrent of hostility, a feeling that the world is against them and does not understand them or take them seriously. With mania as with depression there may be marked irrationality of thought, leading to delusions of persecution or of grandeur and hallucinatory voices which may reinforce their disordered but strongly held inner convictions. Sedative drugs nowadays can calm even the most extreme forms of mania but in earlier times exhaustion and death occurred not infrequently.

Alcoholic psychosis
'Albert' is an example of quite a different psychotic process. Albert has been drinking excessively for years. It started soon after he left school and indeed he could remember being drunk

even before that time when he did odd jobs at one of the local pubs. His first regular employment was on a building site and then at lunchtime he used to drink heavily, more heavily than the older men, as well as after work in the evenings. He was soon in trouble with the police, because he was often in arguments which ended in blows. Albert was conditionally discharged on the first occasion, but the episodes were repeated and a period in prison was the outcome. This did nothing to ease his drinking and when released he became even worse. He found it difficult to hold even a menial job as a barman in a pub in a poor area of the city. Later he did the washing up in a hotel kitchen, but even this kind of lowly task he did badly and did not even turn up on time very often.

Albert had long ago lost contact with his family and had made only a few friends, all drinking men like himself. He had rarely approached women, apart from a passing chat in the bar when he was already under the influence of alcohol. This precarious existence, with another spell in gaol and one in a mental hospital from which he ran away, continued until he was about twenty-six years old. But then a new feature emerged. He began to hear a curious thumping noise in his head. It was brief but repeated and usually after a particularly heavy drinking bout. At first Albert paid little attention, but gradually this noise took on the form of a voice which repeated short phrases like 'He's mad', over and over again. He went to a local hospital on one occasion and saw a psychiatrist who made a diagnosis of a psychotic illness known as 'alcoholic hallucinosis'. Albert was unable and unwilling to go into hospital, or in any way to participate in treatment, but by now the voices were more or less continuous, repeating 'Get him, get him, get him'. Albert took refuge in a bombed-out house, barricaded himself in, and set fire to the place at the behest of his hallucinatory voices, or so he related when he was rescued. However he was so badly burned that he died soon afterwards in hospital.

Albert's psychotic illness is a complication of prolonged excessive alcoholic indulgence and, at this point, it would be useful to compare him with the fire-bug, since it demonstrates how different they are. Diagnostic labels may seem to be of little more than academic value, but this is not really the case, for they are often a guide to how the afflicted person can best be helped as well as how to protect society. Albert had hallucinatory voices which went on more or less continuously. He tended to behave abnormally most of the time. The fire-bug on the other hand, as we have seen, can appear perfectly normal to the onlooker even when he is actually between fires on one of his nocturnal sprees. Albert had not started any blazes before, except on one occasion when in a drunken sleep his cigarette had set light to the bedclothes. The fire-bug, on the other hand, usually reveals several bouts of fire-setting in his history. The reason for Albert burning the house down was of course totally irrational, but he was able to say that it was because the voices had told him to do it. He did not have that irresistible desire to kindle fires so characteristic of the fire-bug. In Albert's case, if he had survived, prison would have been pointless; he needed psychiatric treatment. On the other hand the method of dealing with the fire-bug presents difficulties. From the point of view of society, prison would seem appropriate, yet as he suffers from a severe personality disorder this can only be changed by some form of prolonged psychiatric treatment, a point to which we shall return later.

Albert's illness is similar in type to the more usual schizophrenic disturbance, which is unrelated to prolonged taking of alcohol. Indeed, the cause of this severe mental disorder, which may continue for many years, is in doubt, though fortunately with recently introduced drugs and a more open attitude on the part of the general public many people with schizophrenia eventually return to life in the community. The chronic sufferers are characterised by complex 'delusional systems'. The lowly clerk in a government office, for instance, may begin to feel that he is being spied on by a Russian satellite; as the spy ring reveals itself he becomes aware of someone always walking behind him in the street, waiting for him to reveal official secrets. In the crowded bar where he goes to escape he hears whispered messages from across the other side of the room and sees secret signs which he feels refer to him alone although the gestures are not in the least sinister nor intended for him. We can understand how such a delusional system may eventually lead to the sufferer burning down the house at the end of the street, which he has come to think of as the secret headquarters of the spies. In the past, delusional systems have often involved religion, and the church or synagogue have been chosen for attacks by arsonists, though such attacks are by no means exclusively the prerogative of the psychotic fire-setter.

Epilepsy and fire-setting

There are other types of fire-setter encountered by the psychiatrist, although the term 'psychotic' cannot strictly be applied to them. For example there are those people addicted to narcotic drugs who may, following a 'fix' become confused and delirious, and during the time of incomplete contact with reality may either accidentally, or intentionally for 'kicks', start a blaze. The same is true of the alcoholic who develops *delirium tremens*, a condition accompanied by a blurring of the mental processes. In the past, people with epilepsy have been incriminated as firesetters; and certainly people with epilepsy may, after some sorts of seizure, be found wandering about in a confused state. However, careful examination of the psychiatric literature leads one to believe that epileptics very rarely indulge in fire-setting in the period after a fit, or indeed any other form of criminal activity. The reason for this is quite understandable; in their state of mental confusion the ability for organised behaviour necessary for an arson attack is not present. This at least is the view of Lewis and Yarnell, who did not find a single incident of a fire being set by a person after an epileptic fit.

Why are epileptic people blamed? Perhaps because epilepsy carries a stigma. From surveys it appears that the general population is even more prejudiced against people with epilepsy than it is against those with mental illness, and presumably this is the reason why arson attacks have been blamed on epileptic individuals. This is not just a theoretical argument; recently a young patient of mine, who has very infrequent fits and is on a very small dose of anti-convulsant medication, was brought to me by his parents with the story that a fire had occurred at the school and the headmaster had suggested that because the boy had epilepsy and was on drugs he was quite likely to have started the blaze. It seemed highly unlikely since he was hard-working, well-behaved and not in the least disturbed. All that can be done is to educate the public as to the nature of epilepsy and its significance so that such young people are not exposed unnecessarily to prejudice.

Other disorders
There are various other disorders which lead to degeneration of the brain and consequent bizarre mental disturbance. For example, the final stage of untreated syphilis causes a condition known as general paralysis of the insane. It is a severe form of mental deterioration in which odd and grandiose notions occur and in which behaviour becomes uninhibited. These patients sometimes set dangerous if unplanned fires even in mental hospitals. The occurrence in old people of 'hardening of the arteries of the brain' leads to a similar combination of intellectual impairment and in some instances loss of control. One man I knew was seventy-five years old when his wife died. Since that time he had been living on his own and was barely able to make ends meet, or even to feed and dress himself properly, but he insisted on staying in his own house. He was reluctant, perhaps out of pride, to accept help of any sort. One day he set his place on fire and was later found walking in the street, half-naked and with a knife in his hand. His explanation for the fire was that he wanted to 'smoke out' the strange and evil people who had invaded his house, a clearly irrational idea, and one which was quite out of character. His behaviour was probably caused by the combination of a depressive mood following his wife's death and some brain degeneration due to his age.

Legal issues and the problem of treatment
We must now consider how and where these fire-setters can be treated and what legal issues are involved. First of all, accurate psychiatric diagnosis is of crucial importance. As with other disorders which lead the individual repeatedly into conflict with the law, not only must due consideration be given to the protection of society but all the possibilities for treatment and rehabilitation must be examined as well. This applies equally to the sexually perverted who attack young children and to the psychopath who is repeatedly violent. There is always the same dilemma both at the time of sentencing and of release on parole if the court has given an indeterminate sentence. Some psychiatrists, particularly those who believe strongly in rehabilitation, regard an indeterminate sentence with psychiatric help as the best approach to the chronic offender with a severe personality

disorder. The absence of a fixed span of time may act as a motivating factor aiding treatment, and is thus in this way at least preferable to a prolonged definite sentence. However, with the indeterminate period there is the problem of the possibility of mistakes being made when parole is granted. The recent example of Graham Young comes to mind; he was eventually released from Broadmoor (a special security psychiatric hospital) but almost at once he began poisoning his workmates. Young is an example of a determined, intelligent, psychopathic personality who would probably defy all attempts at rehabilitation, secretly maintaining his interest in toxicology so that when he obtains his release he can at once return to his own perverted pursuits.

Each of the different types of fire-raiser we have examined presents his own particular problem. Let us first consider the psychotic group. Clearcut psychiatric disorders like schizophrenia or depression require psychiatric treatment rather than incarceration as a form of punishment, and this is possible under the current Mental Health Act. The court can arrange for probation of mentally disturbed offenders while they receive treatment, or alternatively for a so-called 'treatment order'. This means that the offender is under compulsion to stay in a mental hospital until the psychiatrists decide that he has recovered from the basic disorder. Clearly it places a strain on the hospital, since the nursing staff must care for and supervise these potentially dangerous people, who may abscond and start another blaze or cause a serious fire in the hospital itself. Locked wards are less frequent now in mental hospitals, but there are certain instances when they are needed and certainly they may be necessary for the fire-setter soon after he has started a series of blazes. Lewis and Yarnell suggest that a period of stabilisation is an important initial step for all arsonists so that they can be observed in detail and assessed fully before a decision is made about their destination and treatment.

It is extremely difficult to decide in a particular case whether treatment has been successful, and quite clear that the ultimate aim of the study of abnormal behaviour is to predict whether or not a particular type of episode will recur. For example, will the fire-setter simply bide his time until he leaves hospital and then start another conflagration? Human behaviour is at the best of times an unpredictable field and scientific study of it is necessarily inexact. Thus it is always difficult to be absolutely certain that treatment has been effective, in part because studies are based on groups rather than individuals and the general applicability of results can be uncertain. Clearly with fire-setting, as with other types of antisocial behaviour, we need more research, so that more appropriate means of psychiatric treatment and rehabilitation can be found.

Another problem arises, as for example in the case of Albert. He was addicted to alcohol and developed a psychosis as a result, but if he had survived the fire he started he would most certainly have required psychiatric treatment. Where could he have been cared for? Intoxication is not regarded as a ground for diminished responsibility and there is a tendency for alcoholics to be sent to gaol where, unfortunately, they may not obtain the appropriate psychiatric treatment simply because the prison service is not geared to it. However, apart from prison or a treatment order in a mental hospital there are other alternatives. Special prisons are to be found with psychiatric facilities available into which people with indeterminate sentences can be placed. The one at Grendon Underwood is a particular example. It is organised on 'group' lines so that the inmates can meet regularly with the staff to discuss both their own problems and those that arise in the general running of the institution. This type of reality-based therapy is now generally accepted in ordinary psychiatric practice as a most useful method of treatment, particularly for those patients who have personality disorders which lead to antisocial behaviour. The regime at Grendon Underwood is not oppressive and there, though a considerable proportion of the inmates are sent because of aggressive behaviour, violent demonstrations of the kind seen recently in other British prisons have not occurred. A large number of arsonists have been treated at Grendon Underwood and this type of institution has the added advantage that research can be carried out, for example into the inner motivations of certain groups. Some of these results will be discussed in the next chapter.

On balance, the then psychiatrically-oriented prison is perhaps the most suitable environment for the fire-bug. As we have said, his basic problem is a long-standing personality disorder which even with expert and prolonged psychiatric treatment can prove difficult to cure. His pattern of abnormal behaviour has often been repeated and the gratification achieved leads to even more ingrained disorder. Such a disturbance, in which there is an inherent conflict between the individual and society, leads to an unhappy, dissatisfied and totally frustrated person; but even here the prolonged influence of psychotherapy can change the pattern. How? Through constant discussion the fire-setter comes to realise what exactly triggers his explosive behaviour. He can learn to avoid the build-up of internal tensions and to divert his pent-up energies in other directions. Such changes cannot be brought about quickly, and in any case are not likely to occur if the offender is merely locked up in an ordinary prison without any psychiatric help. The fire-setter, like many other seriously disturbed offenders, is not adequately cared for at the present time and all we, as members of the general public, can expect is an increasing number of violent crimes in general and malicious fire-setting in particular, if more investigation and research is not undertaken.

The study of Lewis and Yarnell published in 1951 suggested that neither the regimented punitive régime nor a psychiatric hospital in the usual sense of the word are useful. What they think is required is a special type of institution geared to the problems of fire-raisers. I, personally, would not suggest a centre specifically for the treatment of fire-raisers, but would prefer one in which other similar types of offenders with severe personality disorders could also be cared for. Indeed the present prison at Grendon Underwood is probably an ideal set-up, but obviously a big expansion of facilities of this kind is urgently required.

Chapter Thirteen

The Fire-Lovers

Sex and fire

A log fire burning quietly and flickering in the grate creates a feeling of warmth, pleasure and relaxation. A brilliant pyrotechnic display to commemorate some great event, or merely the annual celebration on Guy Fawkes' night, brings out another view of fire. The spectacular devastation of a warehouse is yet another aspect – its destructive ability. Fire has many faces, and has a life-pervading quality which is comparable to sex. But how far can we take this comparison? The fire-bug, we know, has poor sexual adjustment and he certainly gains a gratification akin to sexual pleasure from his contacts with fire. The pleasure gained is not in the form of an orgasm but nevertheless there is a sudden release of inner pent-up tensions. Important here is the fact that though the fire-setter is aggressive towards property and society as a whole, his actual act – the mere striking of a match – is utterly trivial in comparison with the destructive power realised. All he has to do then is to sit back and let the flames take over. Thus though his initial act is aggressive its main quality is one of passivity. Also, his contacts are with inanimate objects, unlike the sexual pervert who attacks women and children, an act which demands personal confrontation.

There are, however, a few fire-raisers whose experience is more directly sexual – the fire-fetishists. Some are vagrants, some are mentally subnormal, but all are solitary and quite withdrawn from society. For these people frequent ecstatic dreams become a kind of addiction and they gradually feel more and more impelled towards fire-raising as a means of achieving gratification. First the pleasure is derived from seeing pictures of flames while masturbating, then there are experiments with harmless blazes, but the impulse grows and grows so that soon the fetishist is aware of a desire for larger and more destructive fires. As the urge takes on ever-increasing proportions, the desire begins to involve people. Frequently fire-bugs choose empty buildings but this is not the case with the fetishist.

Murderous fires
There were an alarmingly large number of hotel fires occurring in London during 1971 which were clearly the work of a different type of fire-setter, the type who is aware of the murderous power of flames. He knows that he has in his hands a magnificent weapon; indeed he is often exquisitely aware of a forbidden pleasure and this apparently yields particular satisfaction. The usual type of fire-bug picks farm buildings or empty premises, but who could select a hotel in the early hours of the morning without realising that the fire must end in destruction of life? These murderous attacks present a particular menace and indicate that we must look carefully at this special type of fire-lover and see if there is a real analogy between murderers and fire-setters, a view put forward earlier in this century by Gross.

Take the case of Jack the Ripper, who attacked and brutally murdered a whole series of women in the East End of London during the nineteenth century. He did not, as far as can be judged, know his victims, who were prostitutes whom he encountered in dark alleyways. Then, like so many other series of murders and fire-setting episodes, they ended as suddenly as they had started. There are many theories about who Jack the Ripper was and why his crimes ceased. As to the latter point, the most likely explanation is that he committed suicide, murder leading to self-murder.

What is the difference between the murderer and the murderous fire-setter? Jack the Ripper had a personal perverse contact with his victim but the fire-lover does not; he starts the blaze and then lets the flames take over, thus slipping quietly from an aggressive to a passive role, a recurring feature of the fire-lover.

The hijackers
The last few years have seen the emergence of a new crime, that of air piracy. Like the highwaymen of the past and the fire-

lovers of the present, the hijackers are almost all males. There is another parallel here, for the hijackers like the fire-lovers can be separated into four main groups: the criminals who do it for money; the politically committed, seeking anarchy and perhaps political ascendancy; the thrill-seekers, caught up in the 'fashion' element and spurred on by detailed and glorifying newspaper reports; and finally the mentally deranged who very often turn to hijacking because of a bizarre inner logic that accompanies schizophrenia. The similarity with the fire-setter is quite remarkable. Another feature which marks both groups is that they have poor personal relationships, and have not in any sense 'made a mark in life'. Some of the hijackers are failed pilots, which instantly reminds one of those fire-setters who, because of their poor abilities, have been rejected as members of the fire-fighting force. There is another parallel in the two groups' dream fantasies. Before the hijacking, the hijacker has constant dreams of incredible events; the fire-lover, too, according to Stekel, has intense and frequent dreams of magnificent conflagrations. But here the analogy ceases, for the hijacker, like the murderer, is in intense conflict with people. The fire-lover resorts to an impersonal weapon – fire.

From all the information gathered so far, what overall picture have we of the fire-lover, his physical and psychological attributes? First of all he tends to be poorly endowed physically and may even be deformed. In addition, we know that with regard to intelligence he is generally at the lower end of the range and some may even be mentally subnormal. Lauretta Bender, as we have mentioned, noted in her series of young children who set fires that they had marked learning disabilities as well as such features as poor hand/eye co-ordination.

These kinds of brain dysfunction can mean that the child is unable to use his intellect to reach his goals, a feature which can have a serious and damaging effect on personality development. It may be that some fire-setters have an even more marked brain disturbance. Studies by Pribram and his colleagues on monkeys, reported by Fulton, showed that with damage to the frontal and temporal lobes of the brain on both sides, there was a total disregard for hot or other dangerous objects and the animals, though apparently able to see and feel normally, were quite unable to learn to discriminate between the noxious and the harmless. It is conceivable that a variation of this pattern may be present in some fire-setters so that the usual learning process about the risks of fire does not occur. Not only do they not understand the dangers, they also come to enjoy this particular element, because it makes them rather different from other people, another feature which could lead to disordered personality development.

Environmental factors in early life are of particular importance in shaping the personality. Again Lauretta Bender noted that not only did her child fire-raisers come from disturbed homes but the degrees of disturbance were much greater than in some of the other groups with disordered behaviour. A factor was the complete absence of one or other parent, especially father, as well as an inconsistent pattern of mothering. All these features tend to produce, as Vandersall and Weiner point out, a sense of exclusion, loneliness and unfulfilled dependency needs: 'While there are no certain unconscious conflicts around specific issues the one consistent factor represented in all the fires was at least temporary breakdown of control in the child.' Because they lived in poor circumstances in which neither their physical nor their psychological needs were fulfilled, these children came to rely entirely on fantasy and these fantasies turned later to ones of violence towards their parents.

Many authors have emphasised that no certain pattern of personality or behaviour characterises the fire-setter. However, like other groups of antisocial offenders, such as those who make sexual attacks on small children, they show certain kinds of personality pattern. There is, first, the immature person, tied to infantile objects. Secondly, there is the so-called disordered personality, individuals who have regressed to primitive forms of sexual behaviour and are unable or unwilling to accept the usual social controls. Finally there is the mentally disturbed and subnormal group who are insufficiently endowed to mature to adult life. As Tennent observes, people with these sorts of personality disorders encounter problems in handling their various inner feelings, none more acute than their aggressive impulses.

Aggression is viewed by different schools of thought in varying ways. It can be seen as an inherent part of human nature, the view of Lorenz, or as a learned response of the person to his environment. Perhaps this latter is the more helpful way of approaching aggression because it suggests the possibility of change and amelioration. In the opinion of Berkowitz, aggression is a response to frustration; aggressive acts therefore lead to the discharge of inner tension and pent-up energies, the type of pattern that we have seen clearly in relation to the group we have called fire-bugs. As we have observed, in Northern Ireland there is also a relationship between aggression, depression and suicide, a topic explained by Kendall.

We will return to this point later, but first of all we must consider some detailed studies carried out on arsonists in the special institutions such as Broadmoor and Grendon Underwood. These are careful reports on male arsonists by McKerracher and Dacre, Hurley and Monahan, and on their female equivalents by Tennent and colleagues. The marked disturbance in family relationships is confirmed by the fact that in one series only ten out of fifty fire-setters came from complete parental homes without any marital problems. The separation of children from parents occurred more frequently in arsonists than in a control group of the criminal population, that is inmates in the same institution who had not turned to the use of fire. On the basis of these studies it is possible to test the validity of the views of psychoanalytic schools that the early years of life are a critical time for the establishment of personal and sexual orientation in life.

Tennent and colleagues, in an investigation involving fifty-six women fire-setters, enquired into sexual development, and this information was once again compared with a control population. Arsonists tended to be more promiscuous and to have convictions for prostitution; fewer were married compared to the control group. In addition, their expressed attitudes on sexual problems showed that, like the men, they had difficulties. It does seem to be true then, that fire-setters as a group, whether male or female, have difficulties over sexual adjustment and this could be related to their early disturbed childhood background. These investigations support the views of Vandersall and Weiner on the question of social adjustment. Hurley and Monahan reported that seventy-five of their sample showed problems of isolation, inability to make friends, fear of involvement and social distrust. Like many other delinquent individuals they tend to come from large families. This may seem rather surprising as one might have expected that there would have been considerable interaction between brothers and sisters. But it is also reasonable to propose that the large number of children meant that the individual child received inadequate attention from his parents, which in turn could lead to increased rivalry between the various members of the family in order to gain recognition from the parents.

The detailed studies mentioned above also allow us to examine the question of ability to deal with explosive and violent impulses. It was found that there was a greater tendency for arsonists to damage property in ways other than by fire, and less likelihood of offences against people, emphasising the earlier point that the fire-setter tends to avoid face-to-face contact. Various personality tests were applied to these groups of arsonists and in summary the results would suggest that they have more difficulty in getting rid of their aggressive feelings than other offenders. This would be compatible with the view that fire-setters experience a build-up of aggressive feelings which they have not learnt to dissipate. Suddenly they lose control; they rush off and kindle a fire and this in itself releases their tensions and they can go to bed and sleep. But there is still the question to be answered, 'Why fire?'

Why fire?
Why is it that children and adults who have disordered backgrounds and personalities often choose fire as a means of disrupting society? Are there alternative methods they could employ? Water seems a possibility, but its destructive power is very limited compared with fire, nor does it perpetuate itself, expanding under its own volition. Fire also has a magical quality and is quite spectacular in its effects. Lauretta Bender was puzzled as to why fire was chosen by her young patients and in detailed case histories she attempted to identify events which could have fixated the children's ideas on fire. She wondered if religious teachings of destruction and purification by fire had been responsible but, at least in past decades, this type of teaching was so widespread that it can hardly be cited as an important factor. She did find that one boy fire-setter had, at the age of eight, been kicked by a fireman and apparently he had sworn to avenge himself of this act. Another of her patients had been burnt by a lamp when

she was very young and another had been impressed by the spectacular activity which resulted when a conflagration broke out in the neighbourhood.

These particular contacts with fire do not seem especially impressive but obviously they might, as it were, be the seeds which fall on fertile ground to be germinated by some later event. This view would be borne out by the story of the pyromaniac given in detail by Lewis and Yarnell, from which I will now recount the important points.

At the age of four or five 'John' had played at burning effigies of his sister, though whether this is fantasy or not is unknown. Two years later he walked on a fire hose and was scolded by a fireman; the importance of the event was perhaps emphasised in his mind by his mother retaliating, asking the fireman why he had picked on such a young child. Already at this young age, however, he was beginning to show an interest in fire, investigating the ruins after blazes had been extinguished.

At the age of ten John had a toy fire engine with a water tower; he experimented by setting a crate on fire and became alarmed when there was not enough water to put it out. Two years later when he was locked in the bathroom there was a blaze in the hall of his home. It caused much excitement and the family were worried about his safety. A year or two afterwards he built fires in the street and roasted potatoes on them but this activity, which might have passed as 'normal', was not nearly so innocent as it appeared; on at least one occasion he threw a handful of bullets on to the bonfire and waited for the explosions. He seemed indifferent to the danger to himself and others. Such a lack of concern appears to be quite usual in many fire-lovers.

At about this period in his life he started collecting sketches of fire engines; he also kindled a fire in a pram in a hallway, then wheeled the pram into the street and called the fire brigade.

Between the years of fifteen and sixteen he lived near a fire station and became intensely interested in the movements of fire engines, watching them from the roof of his house. At about this time too he began experimenting with explosive mixtures.

There now emerges another thread in the story, the occurrence of perverse sexual practices. He became involved in various homosexual pursuits and secretly practised transvestism, using the clothes which had belonged to his sister who had died of tuberculosis. He also became sexually attracted to firemen, and was intensely interested by stories, told to him by a female acquaintance, of certain perverse sexual activities of the local fire chief. In this we see the beginning of his overt association of fire with sex. We see too that the pyromaniac-to-be, rather like the alcoholic with alcohol, is involved early in his life with fire.

Fire-setters, as already observed, cannot control their aggressive impulses; we have also seen that in some cases fire has somehow gained a sexual colouring. But why on a particular day at a certain time does the fire-lover first spring into action? One can understand that there has been the preliminary build up of inner tension, which in ordinary circumstances would be released by sexual or other activity but which the fire-setter is incapable of. There then may occur an event such as dismissal at work which leads him to feel insecure. How is he to deal with these feelings? His limited resources and abilities make a rational approach to the problem difficult. Then the fantasy of fire and its destructive powers, buried in the mind for years, emerge almost like a pantomime genie. As a result, the workshop from which he was dismissed is set ablaze.

Another trigger may be the fire-bug's relationship with women. He gets married, but alas he is impotent on his honeymoon and he rushes out to achieve satisfaction by starting a fire. A manic-depressive fire-setter has severe and sudden swings of mood, changes which may or may not have external precipitants but whatever the cause either in excitement or utter depression a blaze is kindled. There is in these people a delicate balance between the aggressive impulses and the conscious mechanisms which control them. The equilibrium can readily be disturbed so that the 'evil' takes over, if one wishes to use the analogy of Dr Jekyll and Mr Hyde. But this is not the whole story, because the fire-setter often encourages this loss of control over his inner impulses by drinking heavily or taking drugs such as LSD. These and other drugs decrease the ability to control violent and antisocial tendencies and thus lead to fire-setting.

From all that has gone before there clearly emerges the fact that fire is still untamed. Indeed there is a strong suggestion, in spite of improvements in fire-taming technology, that it is gaining the upper hand. It is estimated that by 1990 the number of fires will have at least doubled and may even have trebled. Is the fire-raiser alone responsible for this ever-increasing number of conflagrations? No. Although he is important he is certainly not the only factor.

Another factor contributing to this escalation is the use of synthetic materials rather than traditional ones for furnishings in homes, offices and hotels. If a smouldering cigarette end is inadvertently left on the cushion of a chair covered with plastic and stuffed with synthetic rubber it burns much more quickly than if natural fabrics and padding were employed. And synthetic materials give off intensely toxic vapours which rapidly cause asphyxiation of the victims. Another hazard is the television set, which may blaze up in the middle of the night long after it has been switched off. Again, the main problem here is the highly irritant gases released, often killing the unfortunates in their beds. Polystyrene tiles are now frequently used in homes to cover the ceilings; if a fire begins then these tiles not only catch fire but melt and drip to the floor, thus spreading the flames rapidly.

A further factor appears to be the total lack of concern which many people today exhibit towards fire. They are now no longer experienced in the dangers of naked flames and seem to have little understanding of the damage they can do. Presumably our forefathers, who were constantly exposed to open fires rather than central heating, had no need to be reminded how quickly they could get out of control.

There is also the special problem of the ever-increasing incidence of deliberate fire-setting; of the larger fires, causing damage over £10,000, this is now the major single cause. And as we have seen, there also tends to be under-reporting of such incidents so that the total is in all probability even greater than we realise. The malicious fire-raiser may start blazes all over the country or he may work within a fairly limited area, but whatever his style it is often difficult to determine whether two particular conflagrations were started by the same person. As we know from the case of Leopold Harris, it was the painstaking work of Mr Crocker, collating a mass of apparently dissimilar material, that brought Harris to justice. We now have facilities which make the processing of such data much easier – the computer. If details of all fires from the whole of the country were fed into a computer it could be programmed to indicate whether there were similarities between two blazes which may be separated by miles. This could be done by feeding in facts about the time of day, the means of ignition, where in the building the fire started, and things of this kind. On the basis of previous fire experience which could be programmed into the computer any individual blaze could be compared with all others very quickly. Similarities between conflagrations could be established, and the computer could also indicate whether a particular incident had suspicious features or not. Such a data bank would prove invaluable in alerting the authorities very early after an incident that it was malicious in origin and as a result the offender would be more likely to be caught.

Now what of the deliberate fire-setter himself, assuming he is found? Obviously psychiatric assessment is necessary, and if he is a 'repeater' then clearly he will need to be confined, either in a hospital or in a prison where he has access to psychiatric facilities. This incarceration will have several important effects. It will clearly prevent him from setting fires for a period, and secondly in such an environment he can be studied carefully to find out what features in his personality and environment led to his fire-setting.

And the effects of treatment can also be assessed, for on leaving the institution on parole or at the end of his sentence, his subsequent behaviour can be watched. It is only by vigilance in detecting the malicious fire-setter, by research into the background causes, and the determining of suitable methods of treatment, that we can conquer this particular and ever-increasing problem.

Bibliography

GENERAL

Battle, B. P. and Weston, P. B., *Arson: a Handbook of Detection and Investigation*, Greenberg, New York, 1954.
Gilbert, K. R., *Fire Engines*, HMSO, London, 1966.
Gilbert, K. R., *Fire Fighting Appliances*, HMSO, London, 1969.
Kirk, Paul L., *Fire Investigation Including Fire Related Phenomena: Arson, Explosion, Asphyxiation*, John Wiley, London, 1969.
Lewis, N. D. C. and Yarnell, H., *Pathological Firesetting*. Nervous and Mental Diseases Monograph No. 83, New York, 1951.
O'Dea, W. T., *Making Fire*, HMSO, London, 1964.
Topp, D. O., 'Fire as a symbol and as a weapon of death', *Medicine, Science and the Law*, 13, 79, 1973.

PART ONE: THE BACKGROUND

Chapter 1: The Many Faces of Fire
Bachélard, Gaston, *The Psychoanalysis of Fire* (translated by Alan C. M. Ross), Routledge & Kegan Paul, London, 1964.
Frazer, Sir J. G., *Myths of the Origin of Fire*, Macmillan, London, 1930.
Freud, Sigmund, *The Acquisition and Control of Fire* (vol. XXII, standard edition of collected works), Hogarth Press, London, 1964.
Jung, Carl, *Symbols of Transformation: an Analysis of the Prelude to a Case of Schizophrenia*, Routledge & Kegan Paul, London, 1956.
Paré, Ambroise, *The Apologie and Treatise* (edited Geoffrey Keynes), Falcon Educational Books, London, 1951.

Chapter 2: Fire Out of Control
Ainsworth, W. Harrison, *Old St Pauls*, Collins, London, 1953.
Bell, Walter G., *The Great Fire of London 1666*, Bodley Head, London, 1920.
Dedmon, Emmett, *Fabulous Chicago*, Hamish Hamilton, London, 1954.
Defoe, Daniel, *Journal of the Plague Year*, Dent, London, 1908.
Evelyn, John, *Diary*, as cited by Bell.
Lord, Frank, *Fire Alarm*, Longmans, London, 1957.
Pepys, Samuel, *Diary*, as cited by Bell.
Trevelyan, G. N., *Illustrated Social History of England*, vol. 2, Longman, London, 1950.

Chapter 3: The Destructive Power of Fire
Barnaby, K. C., *Some Ship Disasters and Their Causes*, Hutchinson, London, 1968.
Board of Trade Report on the Accident to Boeing 707-465, G-ARWE, HMSO, London, 1969.
'Fires and Fire Losses classified 1970', *Fire Journal*, published by National Fire Protection Association, USA.
Godson, John, *Unsafe at Any Height*, Anthony Blond, London, 1970.
Gradwohl's Legal Medicine, 2nd edition (edited by Francis E. E. Camps), John Wright, Bristol, 1968.
Polson, C. J., *Essentials of Forensic Medicine*, 2nd edition, Pergamon Press, Oxford, 1965.
Report of Her Majesty's Chief Inspector of Fire Services, HMSO, London, 1970.

Chapter 4: Fire-Raising and Its Investigation
Chambers, E. B., 'Incendiarism: an underestimated danger', *Security Gazette*, 9, 1967.
Fry, J. F. and Le Couteur, B., 'Arson', *Medico-Legal Journal*, 34, 108, 1966.

PART TWO: THE MOTIVATED FIRE-SETTERS

Chapter 5: The Profit Motive
Dearden, Harold, *The Fire Raisers: the Story of Leopold Harris and His Gang*, Heinemann, London, 1934.

Chapter 6: Political Fire-Raising
Hawkes, Nigel, 'How the UN bluffed Britain on napalm', *Observer*, 8 October 1972.

Kenyon, John, 'The guy you don't know', *Observer*, 7 November, 1971.
Lyons, H., 'Depressive illness and aggression in Belfast', *Brit. med. J.*, 1, 342, 1972.
Parker, Tony and Allerton, Robert, *The Courage of His Convictions*, Hutchinson, London, 1962.
Shirer, William L., *The Rise and Fall of the Third Reich*, Secker & Warburg, London, 1961.
Storr, Anthony, *Human Aggression*, Penguin, London, 1968.
Washburn, S. L., 'Conflict in primate society, 1, in *Conflict in Society*, Ciba Foundation and J. A. Churchill, London, 1966.
Watt, Donald, 'The Reichstag fire: Hitler innocent', *Sunday Times*, 1971.
West, D. J., *Murder Followed by Suicide*, Heinemann, London, 1965.

Chapter 7: Ordeal by Fire
Alvarez, A., *The Savage God: a Study of Suicide*, Weidenfeld & Nicolson, London, 1971.
Baez, Joan, *Daybreak*, Panther Books, London, 1971.
Buddhism, translations by Henry Clarke Warren, Atheneum/Harvard University Press, London, 1953.
Hyde-Chambers, F. R., general secretary of the Buddhist Society; personal communication, 1972.
James, W. R. L., 'Suicide by burning', *Medicine, Science and Law*, 6, 48, 1966.
Menninger, C. A., *Man Against Himself*, George Harrap, London, 1938.
Palach, J., 'Full text of letter left by Palach', *The Times*, 12 February 1969.
Ring, Richard, 'Why Jan Palach died', *Observer*, 26 January 1969.
Stengel, Erwin, *Suicide and Attempted Suicide*, Penguin, Harmondsworth, 1964.

PART THREE: THE MOTIVELESS FIRE-SETTERS

Chapter 9: Children as Fire-Setters
Bender, Lauretta, 'Firesetting in Children', in *Aggression Hostility and Anxiety in Children*, Charles C. Thomas, Springfield, Illinois, 1953.
Erikson, E. H., *Childhood and Society*, Penguin, Harmondsworth, 1965.

Kanner, L., *Child Psychiatry*, Charles C. Thomas, Springfield, Illinois, 1957.
Klein, Melanie, *Psychoanalysis of Children*, Hogarth Press, London, 1932.
Macht, L. B. and Mach, J. E., 'The firesetter syndrome', *Psychiatry*, 31, 277, 1968.
Vandersall, T. A., and Wiener, J. M., 'Children who set fires', *Arch. Gen. Psychiat.*, 22, 63, 1970.
Wardrop, F. N., 'Critical reviews : pathological firesetting', *Archives of Criminal Psychodynamics*, 1, 198, 1955.

Chapter 10: The Fire-Bugs
Faecke, Peter, *The Fire Bugs* (translated by Arnold Pomeras), Secker & Warburg, London, 1965.
Krafft-Ebing, Richard, *Impulsive Brandstiftungen*, F. Enke, Stuttgart, 1883.
Lewis, N. D. C. and Yarnell, H., *Pathological Firesetting: Nervous and Mental Diseases*, Monograph No. 82, New York, 1951.
Pearsall, Ronald, *The Worm in the Bud: The World of Victorian Sexuality*, Penguin, Harmondsworth, 1972.
Tennent, T. G., McQuaid, A., Laughnane, T. and Hands, A. J., 'Female Arsonists', *Brit. J. Psychiat.*, 119, 497, 1971.

Chapter 11: Jonathan Martin, Incendiarist of York Minster.
Balston, Thomas, *The Life of Jonathan Martin – Incendiarist of York Minster*, Macmillan, London, 1945.
Marc, C., 'Considerations medico-légales sur la monomanie et particulièrement sur la monomanie incendiaire', *Annals of Public Hygiene and Legal Medicine*, 352, 1833.

Chapter 12: Psychotic Fire-Setters
Aschaffenburg, Gaston, 'Crime and its repression', Criminal Science series 146, 1913.
Crown, Sidney, *Essential Principles of Psychiatry*, Pitman, London, 1970.

Chapter 13: The Fire-Lovers
Berkowitz, Leonard, *Aggression: a Social Psychological Analysis*, McGraw-Hill, New York, 1962.
Fulton, J. F., Pribram, K. H., Stevenson, J. A. F. and Wall, P. D., 'Inter-relations between orbital gyrus, insula, temporal tip and anterior cingulate', *Trans. Am. Neurol. Assoc.*, 175, 1949.

BIBLIOGRAPHY 143

Gross, Hans, *Kriminal-Psychologie* (translated by Horace M. Kallen under title *Criminal Psychology*), Boston, 1911.

Hurley, W. and Monohan, T. M., 'Arson: the criminal and the crime', *Brit. J. Criminol.*, 9, 4, 1969.

Kendell, R. E., 'Relationship between aggression and depression', *Arch. Gen. Psychiat.*, 22, 308, 1970.

Lorenz, K., *On Aggression*, Methuen, London, 1966.

McKerracher, D. W. and Dacre, A. J. I., 'A study of arsonists in a special security hospital', *Brit. J. Psychiat.*, 112, 1151, 1966.

Stekel, W., *Peculiarities of Behaviour*, vol. 2, Bodley Head, London, 1938.

Többen, Heinrich, 'Die Beziehungen zwischer Alkoholisms und Brandstiftungen, Deutsche Zeitschrift für die gesamte gerichlichen, *Medizen*, 23, 235, 1934.

Index

accidental causes of fire, 24, 33, 35
adolescents and fire, 10, 80, 102
aeroplane fires, 26, 27, 31
aggressive behaviour, 61-6, 69, 87, 90, 104, 120, 128, 131, 132, 133, 135
aggressive fires, 61-6
air piracy, see Hijackers
alcoholic psychosis, 120-3, 126
alcoholism, 82, 96, 100
'angry brigade', 62
Arc de Triomphe, 8
arson, 10, 20, 21, 33, 37-40, 43-54, 79-84, 85-92, 93-105, 115, 119-27, 128-37
asphyxiation by fire, 26, 27, 29, 30

babies in fire, 9, 85
Belfast, see Northern Ireland
Bishops Latimer and Ridley, 29, 74
Broadmoor Hospital, 125, 132
Buddhists, 8, 70, 71
burns, 28, 29, 30

carbon monoxide, 30, 31, 32
cauterisation, 6, 7, 13
children and fire, 9, 10, 28, 34, 35, 85-92, 131
Chicago, fire of, 22, 23
Christians, early, 73, 74
cigarettes, 10, 33
cleansing action of fire, 6, 7, 8, 9
control of fire, 3, 10, 12
Constantinople, fire in, 12
crime, see Arson
Crocker, William Charles, 52, 53, 136
Czechoslovakia, 40, 67-74

death by fire, 22-32

death instinct, 73
depression, 62-4, 115, 118, 120, 125, 132, 135
deprived children, 90
dreams of fire, 85, 130
drugs, 83, 135

Easter Sunday, 8
epilepsy, 123
Evelyn, John, 18, 19
exhibitionism, 96, 97

fantasies of fire, 127-8
farm fires, 103
Fawkes, Guy, 10, 55, 56, 106, 128
fire alarms, 99-100
fire balls, 20
fire brigades, 21, 37, 62, 82, 83, 94, 97-100
fire bugs, 10, 80, 81, 84, 93-105, 119, 122, 126, 127, 128
fire fetishists, 128, 129
fire engines, 16, 23, 24, 94, 95, 105
fire-fighting equipment, 16, 21, 23, 24
fire-fighting technology, 3, 16, 136
fire insurance, 21, 22, 44, 46
fire investigation, 33-40
fire loss statistics, 25, 26, 27, 36
fire-raisers, 10, 20, 21, 33, 37-40, 43-54, 79-84, 85-92, 93-105, 115, 119-27, 128-37
Fire Research Station, 33
Fire Protection Associations, 25, 36, 37
firemen, 93, 97-9
firewalking, 8
forensic expert, 31, 32, 37, 74, 75
Freud, Sigmund, 4, 91

INDEX

Germany, 56-60
Grendon Underwood Prison, 126, 127, 132
Gunpowder Plot, 55, 56

Harris, Leopold, 39, 43-54, 80, 136
Hell, 9
hero fire-setters, 82, 89, 94, 96, 101
hijackers, 129, 130
Hippocrates, 6
Hitler, Adolf, 57-60
homosexuals, 60, 87, 89, 134, 135
hospital fire, 81-3
Hubert, Robert, 19, 20
Hus, John, 67, 68

ignition, means of, 34, 35, 38, 43, 50
incendiarism, see also Arson, 33, 34, 37-40, 43-54, 79-84, 88, 106-18
incendiary bombs, 60
intelligence, see also Mental subnormality, 80, 82, 86, 89, 96
insurance, 21, 22, 44, 46
investigation of fires, 33-40
Irish Republic Army, 65

Jack the Ripper, 129
Joan of Arc, 73
Jung, Carl, 4, 5
juvenile fire setters, see Adolescents

Klein, Melanie, 63

legal issues, 124-27
letter-box fires, 35, 36
London Fire Brigade, 21
London, Great Fire of, 14-22
London Salvage Corps, 47, 48
Lubbe, Mirnus van der, 57, 60, 80

magical quality of fire, 29, 69, 70
manic depressive illness, 120, 135
Martin, Jonathan, 106-18, 119
martyrs, 72, 73, 74
masculinity, 91-2
masturbation, 129
masochism, 74-5, 97, 104
medical uses of fire, 6, 13, 14
melancholia, see also Depression, 108
Menai Straits Bridge Fire, 34-5
Menninger, Karl, 73, 74
menstruation, 102
mental subnormality, 80, 88, 131

mental illness, 40, 74, 75, 81, 86, 100, 104, 118, 119, 122, 123, 125-30
Methodists, 107, 108, 109, 111, 112, 116
minimal brain dysfunction, 86, 87
monomania, 116-17
motivated fire-setting, 41-75
motiveless fire-setting, 77-137
murder, 63, 64, 79, 80, 92, 129

napalm, 60, 61
Nero, 12, 18
Northern Ireland, 20, 62-5

oil tanker fires, 27-8
Old Bailey, 19, 43, 53

Palach, Jan, 40, 67-74
passivity in relation to fire, 4, 128
Pepys, Samuel, 14, 16
Phaethon, 5
plastics, 31, 136
political fire-raisers, 39, 40, 55-66, 79
primitive man, 4
profit motive, 43-54
Prometheus, 4, 10, 91
psychiatric disorder, see Mental illness
pyromania, 83, 84, 94, 117, 134, 135

Reichstag fire, 46-60
revenge fires, 79-81, 95
Royston Grange, 28

sadism, 74-5, 97, 104
saints, 72-4
St. Paul's Cathedral, 17, 21
San Francisco, fire of, 13
schizophrenia, 40, 74, 75, 81, 86, 100, 118, 122-5, 130
school fires, 10, 82
self-immolation, 67-75
sexual aspects of fire, 4, 5, 95, 128, 129
ship fires, 27, 28
sterilisation, 7
suicide, 40, 62-5, 67-75, 101, 118
suttee, 72
surgery, see Cauterisation
symbols male and female, 4
symbols of fire, 9, 85, 104

Tasaday tribe, 4
transvestites, 134, 135

United States, 53, 54, 62, 93, 94, 100
Urban communities as fire risks, 3, 12, 97

vagrant fire setters, 81, 100, 101
Vietnam, 60, 61, 69, 70, 71

Whitman, Walt, 93, 94
women fire setters, 101-5

York minster, burning of, 113-16